T0399972

Biological and Pharmacological Properties of the Genus *Moringa*

Biological and Pharmacological Properties of the Genus *Moringa*

Edited by
J. Basilio Heredia and Erick P. Gutiérrez-Grijalva

CRC Press
Taylor & Francis Group
Boca Raton London New York

CRC Press is an imprint of the
Taylor & Francis Group, an **informa** business

First edition published 2022
by CRC Press
2 Park Square, Milton Park, Abingdon, Oxon, OX14 4RN

and by CRC Press
6000 Broken Sound Parkway NW, Suite 300, Boca Raton, FL 33487-2742

Library of Congress Cataloging in Publication Data
A catalog record has been requested for this book

ISBN: 978-0-367-62140-7 (hbk)
ISBN: 978-0-367-62315-9 (pbk)
ISBN: 978-1-003-10886-3 (ebk)

DOI: 10.1201/9781003108863

Typeset in Times
by Newgen Publishing UK

Contents

Preface

This book is intended to be comprehensive and critical literature research of the current state of the art (2010–2021) on the subject of plants of the *Moringa* genus. We found that, to date, there were no comprehensive books regarding studies on all *Moringa* species. We hope that the readers appreciate the information gathered here by specialists in many areas of science, from agriculture, botany, molecular biology, ethnopharmacology, food science, and phytochemistry.

Moringa plants belong to the Moringaceae family, with around 13 species indigenous to India and Pakistan and distributed in many African countries and many parts of the world. Leaves, seeds, bark, and roots from the *Moringa* species have been traditionally used in folk medicine. Many studies have shown that the different parts of *Moringa* plants are a rich source of phytochemicals like phenolic compounds, alkaloids, glucosinolates, isothiocyanates, tannins, and saponins, among others. The evidence shows that regular intake of these molecules is associated with decreased incidence of noncommunicable diseases like cancer, diabetes, metabolic syndrome, cardiovascular diseases, Alzheimer's disease, etc. The ethnopharmacological studies suggest that the bioactive compounds found in *Moringa* species are potential antioxidant, anti-inflammatory, and antidiabetic agents. Our work "Biological and Pharmacological Properties of the Genus *Moringa*" is targeted at scientists, graduate students, and entrepreneurs in the field of agronomy, ethnobotany, and ethnopharmacology of plants, medicinal plants, and *Moringa* species.

In this work, the first three chapters cover agricultural, botanic, and genetic information about *Moringa*. The focus of these comprehensive review chapters is to cover general information on the particularities of *Moringa* crops. Chapters 4 to 11 deal with the phytochemical, ethnopharmacological, and bioactive properties of the chemical constituents of *Moringa* plants and the current and future perspectives on its study.

We appreciate the collaboration of the authors of this book, whose dedication and knowledge allowed this work to be completed and enriched by its multidisciplinary background. We would also like to acknowledge the Consejo Nacional de Ciencia y Tecnología (CONACYT) for the project Cátedras CONACYT from Dr. Erick Paul Gutiérrez-Grijalva and Dr. J. Basilio Heredia.

About the Editors

J. Basilio Heredia is a Senior Researcher at the CONACYT-Centro de Investigación en Alimentación y Desarrollo (Center for Research in Food and Development), at Culiacán, México; since 1996, he has been an active member of the Food Science and Technology Group. He teaches the Phytochemistry and Food Antioxidants courses and collaborates in other classes, such as Doctoral Seminars and Toxicology, Food Technology and Plant Stress Physiology. He has been the major professor of 18 MSc graduate students and six doctoral theses. He has published over 130 research articles, 20 book chapters, two books, and more than 100 conference proceedings at national and international conferences. He is a constant reviewer of journal articles (Journal Citation Reports) and projects for national and international funding agencies. Dr. Heredia is an active member of the CONACYT National Reviewer Board and other academic societies. He has a membership of the Bio-Red, Somech, IFT and has been a member of the Mexican National Research System (SNI) since 2009. His research lines are Science and Technology (Conventional and Emerging) of Functional and Nutraceutical Foods; Secondary Plant Metabolism, Stress and Molecular Signaling (Metabolomics); and Plant Bioresources as Source for Antioxidants and Bioactive Compounds. He has experience in projects with companies in the Food Agro-Industrial Area, and has carried out projects with National and International Funding Agencies, such as Mexican SADER, FOMIX, SE-PYME, UC Mexus, TAMU-Conacyt, PEI-Innovation and other private funds. He collaborates with colleagues from institutions and universities from Mexico, the USA, Canada, and Spain.

Erick Paul Gutiérrez-Grijalva is currently a CONACYT Researcher at the Centro de Investigación en Alimentación y Desarrollo, A.C., which is a Federal Public Research Center. Dr. Gutiérrez-Grijalva gained his PhD degree in 2018 at the Laboratory of Functional Foods and Nutraceuticals evaluating the bioactivity and bioaccessibility of three Mexican oregano species. He has experience in the study of phytochemicals from plant-based foods and medicinal plants, and in the characterization, identification, and bioaccessibility of phytochemicals and their bioactive potential as antioxidant agents against oxidative stress, metabolic syndrome, diabetes, and cancer. His research interests include: polyphenols, phytochemicals, ethnopharmacology, and food science. His research projects include "Nutraceutical Potential of Medicinal Plants", by Cátedras CONACYT, and "Functional genomics applied to the characterization of the antimicrobial effect of Mexican oregano against the pathogenic free-living amoeba *Balamuthia mandrillaris*", by the grant Ciencia de Frontera 2019 given by CONACYT.

List of Contributors

Diana Marcela Arias-Moreno Grupo de Investigación BIOPLASMA, Facultad de Ciencias, Universidad Pedagógica y Tecnológica de Colombia, Tunja, Colombia

Sara Avilés-Gaxiola Centro de Investigación en Alimentación y Desarrollo A.C., Carretera a Eldorado Km. 5.5, Col. Campo El Diez, Culiacán, Sinaloa, México

R. Baeza-Jiménez Laboratorio de Biotecnología y Bioingeniería, Centro de Investigación en Alimentación y Desarrollo, Chihuahua, México

Manuel A. Picos-Salas Centro de Investigación en Alimentación y Desarrollo A.C., Carretera a Eldorado Km. 5.5, Col Campo El Diez, Culiacán, Sinaloa, 80110 México

Luis A. Cabanillas-Bojórquez Centro de Investigación en Alimentación y Desarrollo A.C., Carretera a Eldorado Km. 5.5, Col Campo El Diez, Culiacán, Sinaloa, México

Ramón Ignacio Castillo-López Facultad de Ciencias Químico Biológicas, Universidad Autónoma de Sinaloa, Ciudad Universitaria, Culiacán, Sinaloa, México

Carlos Bell Castro-Tamayo Facultad de Medicina Veterinaria y Zootecnia, Universidad Autónoma de Sinaloa, Culiacán, Sinaloa, México

Laura A. Contreras-Angulo Centro de Investigación en Alimentación y Desarrollo A.C., Carretera a Eldorado Km. 5.5, Col Campo El Diez, Culiacán, Sinaloa, México

Abraham Cruz-Mendívil Cátedras CONACYT-Instituto Politécnico Nacional, CIIDIR Unidad Sinaloa, Departamento de Biotecnología Agrícola, Guasave, Sinaloa, México

J. Abraham Domínguez-Ávila Cátedras CONACYT-Centro de Investigación en Alimentación y Desarrollo, Hermosillo, Sonora, México

Cristina Alicia Elizalde-Romero Centro de Investigación en Alimentación y Desarrollo A.C., Carretera a Eldorado Km. 5.5, Col Campo El Diez, Culiacán, Sinaloa, 80110 México

Alexis Emus-Medina Universidad Autónoma de Sinaloa. Facultad de Medicina Veterinaria y Zootecnia, Blvd. San Ángel 3886, Mercado de Abastos, San Benito, Culiacán 80260, Sinaloa, México

R. Flores-Flores Laboratorio de Biotecnología y Bioingeniería, Centro de Investigación en Alimentación y Desarrollo, Chihuahua, México

Melissa García-Carrasco Centro de Investigación en Alimentación y Desarrollo A.C., Carretera a Eldorado Km. 5.5, Col Campo El Diez, Culiacán, Sinaloa, México

Gustavo A. Gonzalez-Aguilar Coordinación de Alimentos de Origen Vegetal, Centro de Investigación en Alimentación y Desarrollo, Hermosillo, Sonora, 83304, México

Roberto Gutiérrez-Dorado Facultad de Ciencias Químico Biológicas, Universidad Autónoma de Sinaloa, Culiacán, Sinaloa, México

Janet Alejandra Gutiérrez-Uribe
Tecnologico de Monterrey, Av. Eugenio
Garza Sada 2501 Sur, Col. Tecnologico,
Monterrey, N.L., México

Liseth Daniela Jaimes-Rangel
Universidad Francisco de Paula
Santander, Ingeniería Biotecnológica,
San José de Cúcuta, Norte de Santander,
Colombia

**Janitzio Xiomara K. Perales-
Sánchez** Facultad de Ciencias Químico
Biológicas, Universidad Autónoma de
Sinaloa, Culiacán, Sinaloa, México

Leticia X. López-Martínez Laboratorio
de Antioxidantes y Alimentos
Funcionales, CONACYT-Centro
de Investigación en Alimentación y
Desarrollo, Hermosillo, Sonora, México

Luisa Fernanda Madrigales-Reátiga
Facultad de Ciencias Químico
Biológicas, Universidad Autónoma de
Sinaloa, Culiacán, Sinaloa, México

Julio Montes-Ávila Facultad de
Ciencias Químico Biológicas,
Universidad Autónoma de Sinaloa,
Ciudad Universitaria, Culiacán,
Sinaloa, México

Luis A. Montoya-Inzunza Centro
de Investigación en Alimentación y
Desarrollo A.C., Carretera a Eldorado
Km. 5.5, Col Campo El Diez, Culiacán,
Sinaloa, 80110 México

Rosmarbel Morales-Nava Tecnologico
de Monterrey, Av. Eugenio Garza Sada
2501 Sur, Col. Tecnologico, Monterrey,
N.L., México

M. A. Morales-Ovando Facultad
Ciencias de la Nutrición y Alimentos,
Universidad de Ciencias y Artes de
Chiapas, Chiapas, México

Elvia Paulina Pérez-Rivera Facultad
de Ciencias Químico Biológicas,
Universidad Autónoma de Sinaloa.
Ciudad Universitaria, Culiacán,
Sinaloa, México

Manuel Bernal-Millan Centro de
Investigación en Alimentación y
Desarrollo A.C., Carretera a Eldorado
Km. 5.5, Col Campo El Diez, Culiacán,
Sinaloa, México

Jesús José Portillo-Loera Facultad
de Medicina Veterinaria y Zootecnia,
Universidad Autónoma de Sinaloa,
Culiacán, Sinaloa, México

Cuauhtémoc Reyes-Moreno Facultad
de Ciencias Químico Biológicas,
Universidad Autónoma de Sinaloa,
Culiacán, Sinaloa, México

Norma Julieta Salazar-Lopez
Coordinación de Alimentos de Origen
Vegetal, Centro de Investigación en
Alimentación y Desarrollo, Hermosillo,
Sonora, 83304, México

1 The *Moringa* Genus
Botanical and Agricultural Research

Luisa Fernanda Madrigales-Reátiga
Facultad de Ciencias Químico Biológicas,
Universidad Autónoma de Sinaloa,
Culiacán, Sinaloa, México.

Roberto Gutiérrez-Dorado
Facultad de Ciencias Químico Biológicas,
Universidad Autónoma de Sinaloa,
Culiacán, Sinaloa, México.
Corresponding author: robe399@hotmail.com

Janitzio Xiomara K. Perales-Sánchez
Facultad de Ciencias Químico Biológicas,
Universidad Autónoma de Sinaloa,
Culiacán, Sinaloa, México.

Cuauhtémoc Reyes-Moreno
Facultad de Ciencias Químico Biológicas,
Universidad Autónoma de Sinaloa,
Culiacán, Sinaloa, México.

CONTENTS

DOI: 10.1201/9781003108863-1

1.1　INTRODUCTION

The genus *Moringa* belongs to the Moringaceae family, which is known as the 'drumstick' or 'horseradish' family; it comprises 13 species (Rani et al., 2018). *Moringa* species are native to Africa and Asia, but they have been widely distributed in many tropical and subtropical countries worldwide (Mallenakuppe et al., 2019).

The *Moringa* is a fast-growing softwood tree that can reach 10–12 m in height; it has good adaptability to humid and dry climates. Its pinnate leaves and long, woody pods are recognized when ripe as they open in three valves containing globular seeds. Farmers, scientists, and researchers have focused attention on *Moringa* due to its various nutritional and medicinal properties and use of the ornamental plant for animal forage (Lalas and Tsaknis, 2002; Leone et al., 2015). These plants are used to produce biodiesel, biogas, water purification cleaning agents, fertilizers, foliar nutrients, and green manures (Padayachee and Baijnath, 2012; Boukandoul et al., 2018). These uses are attributed to the plant's different parts (fruits, seeds, leaves, flowers, bark, and roots).

Due to the nutritional, medicinal, energy, and agricultural benefits that can be obtained from this plant, it is of great importance to know the botanical characteristics, geographic distribution, and agricultural management of *Moringa* species to obtain the maximum use of all parts of the plant in the different places where it is grown.

1.2　BOTANICAL DESCRIPTION

1.2.1　TAXONOMY

Moringa is the only genus of the family Moringacea, which has been classified taxonomically in different ways, being placed in the Rheodales, Capparales, and Violaceae over the years (Ramachandran et al., 1980). *Moringa*'s relationships with other plant families have long been controversial; this family was placed in the Capparales because of its zygomorphic flowers, the presence of a gynophore, parietal placentation, and lack of endosperm. In the same way, it has been placed in Violacea because of superficial similarities in the pentamerous zygomorphic flowers and embryologic characters. However, despite these associations and similarities with families, *Moringa* remains a Moringaceae family member (Rodman et al., 1996; Shindano and Kasase, 2009); within the order, Capparales includes the cabbage and radish families (Spandana et al., 2016; Velázquez-Zavala et al., 2016; Habtemariam, 2017; Gandji et al., 2018).

1.2.2 MORPHOLOGICAL CHARACTERISTICS

The genus *Moringa*, the only one within the Moringaceae family, includes 13 species [*M. arborea* Verdc., *M. borziana* Mattei, *M. concanensis* Nimmo, *M. drouhardii* Jum., *M. hildebrandtii* Engl., *M. longituba* Engl., *M. oleifera* Lam., *M. ovalifolia* Dinter & Berger, *M. peregrina* (Forssk.) Fiori, *M. pygmaea* Verdc., *M. rivae* Chiov., *M. ruspoliana* Engl., and *M. stenopetala* (Bak. f.) Cuf.] (Olson, 2002a; El-Kheir et al., 2017; Hassanein and Al Soqeer, 2018; Ruiz et al., 2019).

A set of unique traits distinguishes the Moringaceae family, of which large and pinnate leaves that can measure up to 60 cm, divided into leaflets on a rachis, stand out. Leaf glands are borne from a short stem. There are conspicuous pedunculated glands where the petiole intersects the stem. Each pinna intersects the rachis and each leaflet intersects an axis (Olson, 2002a; Abdel-Hameed, 2015; Habtemariam, 2017). Its fruits form a long and woody capsule that at maturity measures up to 50 cm (Olson and Fahey, 2011). *Moringa* seeds have three longitudinal wings in most species with an outer mesotesta of thin-walled cells, varying depending on the species from 5 to 25 cells. The central mesotesta consists of sclereids ranging from 20 to 50 cells in wingless species. The internal mesotesta has a thickness of 10–30 cells depending on the species. Other less-noticeable characteristics of *Moringa* are the hollow style and the anthers with two sporangia or chambers for the poles, rubber ducts in the medulla of the stems, and vascular elements with perforated plates without borders (Olson, 2002b).

Moringaceae is one of the 15 or so families that produce mustard oils (glucosinolates) like the following families: mustard family (Brassicaceae), the caper family (Capparacea), the nasturtium family (Tropaeolaceae), and the papaya family (Caricaceae) (Palada et al., 2019). The closest family to Moringaceae is Caricaceae, that of papaya; both share several characteristics in terms of life form, wood, bark, leaves, floral ontogeny, merosity, and flower sexuality, showing that Caricaceae and Moringaceae are sister taxa (Olson, 2002b; Olson and Fahey, 2011; Azza, 2014). Among the similar characteristics reported by Olson (2002b), the following stand out: bottle tree, paratracheal axial parenchyma dominating xylem, thick bark, palmately compound leaves, palmate venation, leaf glands, contorted growth of flowers, five sepals, five petals, ten androecial elements, bisexual flowers, and trichomes on filament bases.

Although the *Moringa* genus is small for its size, one of the most phenotypically varied groups of angiosperms includes huge "bottle trees" to tiny tuberous shrubs (Olson, 2002a; Shindano and Kasase, 2009). Different groups of species within *Moringa* have been reported based on their floral and leaf morphology, palynology, habit, and wood anatomy (Olson and Carlquist, 2001; Abdel-Hameed, 2015). Table 1.1 shows the classification according to its floral morphology, in which the genus is divided into three sections: *Moringa* (bilateral symmetry with a short hypanthium) consists of eight species of tuberous shrubs and sarcorhizal trees with thick and fleshy tuberous roots;

TABLE 1.1
Moringa **Species Groups**

Species by section (Verdcourt, 1986)	Species by habit (Olson and Carlquist, 2001)
Moringa: *M. arborea*, *M. borziana*, *M. concanensis*, *M. oleifera*, *M. peregrina*, *M. pygmaea*, *M. rivae*, *M. ruspoliana*. Flowers irregular, perigynous; receptacle short; seeds winged or almost winged	Bottle trees: *M. drouhardii*, *M. hildebrandtii*, *M. ovalifolia*, *M. stenopetala*. It is characterized by massive trunks and swollen roots, where it stores large amounts of water
Dysmoringa: *M. longituba*. Flowers irregular, perigynous; receptacle tubular, long; seeds winged	Sarcorhizal trees: *M. arborea*, *M. ruspoliana*. They are characterized by a slender trunk with tough, smooth bark and thick, soft, fleshy, contorted roots
Donaldsonia: *M. drouhardii*, *M. hildebrandtii*, *M. ovalifolia*, *M. stenopetala*. Flowers regular or nearly so, hypogynous; receptacle short; seeds very strongly winged or not winged	Slender trees: *M. concanensis*, *M. oleifera*, *M. peregrina*. Slender trunks and tough, fibrous roots
	Tuberous shrubs: *M. borziana*, *M. longituba*, *M. pygmaea*, *M. rivae*. Slender, often deciduous shoots, with huge tuberous roots

M. arborea, *M. borziana*, *M. concanensis*, *M. oleifera*, *M. peregrina*, *M. pygmaea*, *M. rivae*, and *M. ruspoliana*. A second section is Dysmoringa (bilateral symmetry with a long hypanthium), known as the "tuberous clade", which includes *M. longituba*. The third group is Donaldsonia or the "bottle tree" group (radial symmetry), consisting of four species, *M. drouhardii*, *M. hildebrandtii*, *M. ovalifolia*, and *M. stenopetala*, with swollen trunks and radially symmetrical flowers (Verdcourt, 1986; Olson, 2002a; Abdel-Hameed, 2015). *Moringa* is divided into four habit classes based on its general appearance: bottle trees, sarcorhizal trees, slender trees, and tuberous shrubs.

The bottle trees, slender trees, and tuberous shrubs display anatomical homogeneity, while the two species of sarcorhizal trees show marked anatomical differences (Table 1.1) (Olson and Carlquist, 2001; Shindano and Kasase, 2009).

The 13 species of *Moringa* contained in each of the classifications present different morphological characteristics. Table 1.2 shows the following parameters: tree height, trunk diameter, branch length, leaf length, leaflet shape, number of leaflets, number of seeds, seed length, flowers, fruit, seed shape, seed texture, seed color, and wing form for five species. *Moringa* trees show marked differences in their morphology; some of its 13 species' main characteristics are described below.

1.2.2.1 *Moringa arborea* Verdc.

Moringa arborea is a shrub or tree up to 15 m tall, bark smooth, gray, and slender; this genus has its main characteristic large flowers and seeds, glabrous stems, and a close resemblance to *M. rivae* (Verdcourt, 1985; Olson and Carlquist, 2001). The flowers are cream-colored, with a pale pink calyx and deeper pink at the base, in subterminal

TABLE 1.2
Morphological Characteristics of Five *Moringa* Species

Parameter	*M. concanensis*	*M. oleifera*	*M. ovalifolia*	*M. peregrina*	*M. stenopetala*
Tree height (m)	2.13–2.43	6–12	4–8.5	6–15	6–12
Trunk diameter (cm)	15	20–45	N/A	170–190	60
Branch length (cm)	45	30–75	N/A	250	25–67
Leaf length (cm)	10–20	25–52	30–70	22–75	33–50
Leaflet shape	Linear	Oboordate	Oval	Linear	Elliptic
No. leaflets/leaf	3	9	4–10	6	7
No. seeds/pods	N/A	12–18	N/A	11–14	9–11
Seed length (cm)	3.5–4	0.9–1.9	1.3–4	1.2–2	2.5–3.5
Flowers	Yellow with red or pink veins	Fragrant, yellowish-white	Small, white	Fragrant, creamy-white	White
Fruit	Straight capsules	Pod-like capsule	Pod-like capsule	Elongate capsule	3-valve elongated capsule
Seed shape	Angled	Round	Winged	Globus to ovate	Elliptical
Seed texture	N/A	Rough	Smooth	Smooth	Rough
Seed color	White or pale yellow	Brown	Brown	Brown	Creamy
Wing form	Winged, very thin, hyaline	Papery	N/A	N/A	Papery
References	Balamurugan and Balakrishnan, 2013; Santhi and Sengottuvel, 2016; Tahir et al., 2020	Azza, 2014; El-Kheir et al., 2017	Habtemariam, 2017; Hausiku et al., 2020	Azza, 2014; El-Kheir et al., 2017	Azza, 2014; El-Kheir et al., 2017

N/A, not applicable

inflorescences up to 15 cm long. Young pods are greenish becoming gray-brown on dehiscence, and seeds are greenish with hyaline wings (Padayachee and Baijnath, 2012; Habtemariam, 2017).

1.2.2.2 *Moringa borziana* Mattei

Moringa borziana is a woody herb or small shrub that can grow up to 5 m high. The leaves have pinnate, glabrous, and stipitate appearances; as for the leaflets, their color varies from pale green to yellow and shape elliptical to obovate (Olson and Carlquist, 2001). The flowers are considered zygomorphic, yellowish-red and spatulate. The fruits are waxy, purplish-brown; the seeds have three conspicuous wings and are 3.8 cm long (Padayachee and Baijnath, 2012; Habtemariam, 2017). The plant's characteristic feature is being tuberous, giving rise to new shoots after periodic dieback of the aerial parts every few years (Habtemariam, 2017).

1.2.2.3 *Moringa concanensis* Nimmo

Moringa concanensis is a small, evergreen tree that is glabrous except for the young leaves and inflorescences, with a spreading crown, up to 2.44 m tall. The leaves are pinnate, obovate, caducous, 45 cm long, and have three linear leaflets (Olson and Carlquist, 2001; Anbazhakan et al., 2007; Manzoor et al., 2007). This species has linear pods of 30–45 cm length and which are sharply three-angled. A distinctive layer of very furrowed bark covers the central trunk. Flowers are large, white, hermaphrodite with petals yellow, veined with red, and oblong in shape. It shows 5 fertile stamens and 4–5 staminodes (Padayachee and Baijnath, 2012; Balamurugan and Balakrishnan, 2013). The capsules are straight, actively triquetrous, slightly constricted between the seeds, white or pale yellow, and three-angled. The horseradish odor of *M. concanensis* is more intense than *M. oleifera* (Santhi and Sengottuvel, 2016; Boukandoul et al., 2018; Tahir et al., 2020).

1.2.2.4 *Moringa drouhardii* Jum.

Moringa drouhardii is a small or big deciduous tree 10–18 m tall. Its main feature is the swollen stem (Arora et al., 2013; Palada et al., 2019), short branches, often unbranched, whitish bark with resin, alternate tripinnate leaves. The flowers are characterized by being bisexual, regular, and ovate with yellowish-white petals. The fruits are 30–50 cm long and are elongated trigonal capsules with a beak and glabrous. The white seeds, trigone to ovoid, do not have wings (Olson and Carlquist, 2001; Padayachee and Baijnath, 2012; Habtemariam, 2017).

1.2.2.5 *Moringa hildebrandtii* Engl.

Moringa hildebrandtii is a deciduous tree that can reach up to 25 m high, with a trunk bloated by massive water-storing (Olson and Razafimandimbison, 2000). Leaves are compound and tripinnate; they can grow up to one meter long. The flowers are whitish and small, borne in large sprays (Olson and Carlquist, 2001; Arora et al., 2013). The fruits are spindle-shaped capsules that measure between 450 and 650 mm long and are constricted between the seeds, which are pale brown, ovoid trigonous, winged, and 35–40 mm long (Padayachee and Baijnath, 2012; Habtemariam, 2017).

1.2.2.6 *Moringa longituba* Engl.

Moringa longituba is a shrub; it ranges in height from 2 to 6 m with smooth, pale gray bark. The leaves are pinnate of 2–3, young densely pubescent and later glabrescent; the leaflets are oblong, elliptic, or obovate (Padayachee and Baijnath, 2012; Habtemariam, 2017). The flowers are bilateral and symmetrical; this species' main characteristics are the bright red flowers or the petal and sepal bases with a long, tubular hypanthium. The fruits are purple-brown with bloom, and the seeds are 2–3 cm long and winged. It usually has a large tuber at great depth (Arora et al., 2013).

1.2.2.7 *Moringa oleifera* Lam.

Among the 13 species of trees and shrubs of the genus *Moringa*, *M. oleifera* is the best known. The regional names by which this species is commonly known are drumstick tree, kelor, murungai kaai, saijhan, and benzolive; it is a fast-growing, evergreen, deciduous, medium-sized perennial tree, ranging in height from 10 to 12 m; it has tuberous roots, whitish bark, and is surrounded by thick cork, soft, spongy wood, a short trunk, and slender branches (Padayachee and Baijnath, 2012; Azza, 2014; Leone et al., 2015). Young shoots have purplish or greenish-white bark. The leaves up to 60 cm long are alternate, twice- or thrice-pinnate, and spirally arranged with

FIGURE 1.2 *Moringa* tree and its parts. (A) *Moringa* tree; (B) large, pinnate leaves; (C) fruit, it is a light capsule, opening into three parts or valves; (D) yellowish-white fragrant flowers.

articulated primary, secondary, and tertiary rachis. With an average of nine leaflets, leaflet shape is oboordate (Velázquez-Zavala et al., 2016; El-Kheir et al., 2017; Hassanein and Al-Soqeer, 2018). The flowers are white or cream, 2.5 cm wide, and pleasantly fragrant. The fruits are pod-shaped capsules that are born singly or in pairs. They are light green, thin, tender, firm, and of fine wood. The matured fruit consists of a 20–45 cm hanging capsule that contains between 15 and 20 seeds. The dark brown seeds are spherical, 7–8 mm in diameter, with four paper-like wings (Figure 1.2) (Rani et al., 2018; Mallenakuppe et al., 2019; Ruiz et al., 2019; Tahir et al., 2020).

1.2.2.8 *Moringa ovalifolia* Dinter & Berger

Moringa ovalifolia, known as phantom or ghost tree, is an erect, deciduous tree with a swollen stem that grows up to 7 m high (Olson and Carlquist, 2001). It has a smooth bark that is pale gray and shiny with resin. The leaves are compound and alternate of light green color, reaching a length of up to 60 cm with approximately nine leaflets of up to 25 mm long and white flowers sprouting from branched axillary sprays (Hausiku et al., 2020; Tahir et al., 2020), that can grow up to 3 mm in size with 4–5 petals. Pod-like three-sided fruits, brown color, of about 400 mm long (Padayachee and Baijnath, 2012; Habtemariam, 2017).

1.2.2.9 *Moringa peregrine* (Forssk.) Fiori

Moringa peregrina is an extremely fast-growing tree or shrub with an estimated 3–10 m in height in just 10 months from a seedling (Osman and Abohassan, 2012; Padayachee and Baijnath, 2012; Azza, 2014). It has grayish-green bark; long, pinnate, and alternate leaves with around three pairs of a leaflet, with linear in shape; and yellowish-white to fragrant pink flowers. The fruits are shaped like elongated capsules, which end in a beak, are glabrous and slightly narrow between the seeds (El-Kheir et al., 2017; Habtemariam, 2017; Boukandoul et al., 2018). The shape of the seed is globose to ovoid or trigone. It has high medicinal value, which is why this plant is widely used; it is among the best-known Arab medicinal plants and has many other uses (Hassanein and Al-Soqeer, 2018; Tahir et al., 2020).

1.2.2.10 *Moringa pygmaea* Verdc.

Moringa pygmaea is a delicate, tuberous shrub or herb with stems up to 15 cm tall from the large tuberous rootstock (Olson and Carlquist, 2001). The leaves on this plant are pinnate, glabrous with 3–4 pairs, and the leaflets with 3–5 per pinna are obovate and yellowish-green. The dull yellow or purplish-brown flowers are bisexual and borne in axillary panicles, sepals 0.9–1.2 cm long, ciliate at the apex, and petals 1.2–1.3 cm long (Padayachee and Baijnath, 2012; Habtemariam, 2017). The fruits are capsules, ribbed, and sometimes appear with an elongated beak, and the seeds are 3-winged or wingless, without endosperm (Arora et al., 2013).

1.2.2.11 *Moringa rivae* Chiov.

Moringa rivae is one of the less-investigated *Moringa* species; it is a small, slender shrub or tree. The leaves are characterized by being alternate, pinnate, and the leaflets are oblong to elliptical and glabrous (Olson and Carlquist, 2001). The flowers have a honey odor and are yellowish or reddish, and the fruits are hanging pods with 9 ribs, 30–45 cm long, and tomentose when young. Seeds are embedded in the valves' pits and are three-angled, winged, blackish, and rounded (Padayachee and Baijnath, 2012; Habtemariam, 2017).

1.2.2.12 *Moringa ruspoliana* Engl.

Moringa ruspoliana is morphologically easy to distinguish among the Moringaceae family. This species' main feature is its simple pinnate leaves and large flowers (Verdcourt, 1985; Arora et al., 2013). The leaflets are the largest (reaching almost 15 cm in diameter), thick, and resistant. The flowers are pink with green bases (Olson and Carlquist, 2001; Padayachee and Baijnath, 2012; Habtemariam, 2017).

1.2.2.13 *Moringa stenopetala* (Bak. f.) Cuf.

Moringa stenopetala is a small tree, bark whitish, pale gray, smooth, up to 10 m tall with a swollen, bottle-shaped trunk and alternate, pinnate leaves, up to 55 cm long with elliptical to ovate leaflets (Bosch, 2004; Arora et al., 2013). The inflorescence is dense, with many-flowered panicles up to 60 cm long, which are bisexual and regular; with free sepals and cream flushed pink petals; linear–oblong, 8–10 mm long, five stamens, and four filaments. The fruits are elongate 3-valved capsules, 20–50 cm long; grooved, twisted when young, later straight, reddish with grayish bloom, and many-seeded. Seeds are elliptical–trigone in shape, 6–9 cm long, with three thin wings (Padayachee and Baijnath, 2012; Habtemariam, 2017).

1.2.3 *MORINGA OLEIFERA* LAM

Moringa oleifera is the species best known and most researched in the genus. It is considered a 'miracle tree' for its many uses and for all parts of the plant being useful for humans, which has been the subject of attention from researchers, farmers, and scientists, among others (Palada et al., 2019). *Moringa oleifera* is commonly known by names such as drumstick tree, kelor, murungai kaai, saijhan, and benzolive (Shindano and Kasase, 2009; Padayachee and Baijnath, et al., 2012); it is a rapidly growing tree, drought-resistant, native to northwestern India and Pakistan. The native range is usually the foothills of the Himalayas in northwest India (Leone et al., 2015; Tahir et al., 2020), and widely cultivated in tropical and subtropical areas, typically grows in a semi-dry desert or tropical soils; it has now become naturalized in Afghanistan, Florida, and East and West Africa (Paliwal et al., 2011).

Each part of the *Moringa* tree is associated with the presence of at least one benefit, but the leaves are the part most used globally (Mallenakuppe et al., 2019). Various studies over the years have endorsed and recognized the medicinal uses of *M. oleifera* (Rani et al., 2018); used for treating fevers, dysentery, asthma, dental decay (gum); warts (seeds) (Padayachee and Baijnath, 2012); diarrhea, dysentery, colitis, sores, skin infection, anemia, scrapes, rashes, signs of aging, antibacterial, antimalarial, hypertension, diabetes, arthritis, cardiac stimulants, diseases of the skin, typhoid fevers, malaria, swellings, parasitic diseases, cuts, contraceptive remedy, genito-urinary ailments, boost the immune system, elicit lactation (leaves) (Jaiswal et al., 2009; Kasolo et al., 2010; Abe and Ohtani, 2013; Yabesh et al., 2014); gout, acute rheumatism (oil and flowers); toothache, anthelmintic, antiparalytic (roots) (Sivasankari et al., 2014; Rani et al., 2018); aiding digestion, stomach pain, poor vision, ulcers, hypertension, joint pain, anemia, diabetes (barks) (Yabesh et al., 2014).

Moringa oleifera seeds are an effective water-clarifying agent for various colloidal suspensions, used mainly by the rural population living in extreme poverty (Arora et al., 2013). *Moringa oleifera* also exerts many pharmacological activities such as anticancer, antioxidant, anti-inflammatory, and immunomodulatory (Sudha et al., 2010; Castillo et al., 2016, 2017; Dhakad et al., 2019; Chhikara et al., 2020; León-Lopez et al., 2020). For all this, *M. oleifera* is considered a very valuable plant, and is under constant investigation, being a functional and highly useful food for humans.

1.3 OCCURRENCE AND DISTRIBUTION

The monogeneric Moringaceae family has 13 species of tropical and subtropical flowering trees (Ramachandran et al., 1980; Leone et al., 2015; Boukandoul et al., 2018; Özcan, 2018; Tahir et al., 2020). It has been recognized for hundreds of years as a remedy for various diseases; known for treating over 300 health-related problems (Palada et al., 2019); it is presumed that all *Moringa* species are native to India and Africa, but have been distributed into several countries of the tropics over the years (Table 1.3; Figure 1.3).

Moringa oleifera, the most researched species of the genus *Moringa* due to its multiple uses and potential, is originally from the Himalayas and native to north-western India and Pakistan (Shindano and Kasase, 2009; Paliwal et al., 2011; Pandey et al., 2011). This species has attracted attention over the years; since the Ancient Egyptians, who used *Moringa* oil for cosmetic pruposes and in the preparation of the skin, traditional healers have utilized different parts of the *Moringa* tree as traditional

TABLE 1.3
***Moringa* Species Origin and Distribution**

Species	Origin	Distribution
Moringa arborea Verdc.	Kenya[2]	Kenya and Somalia[1]
Moring borziana Mattei	Kenya and Somalia[2]	Kenya and Somalia[1]
Moringa concanensis Nimmo	India[2]	India[1]
Moringa drouhardii Jum.	Madagascar[2]	Madagascar[1]
Moringa hildebrandtii Engl.	Madagascar[2]	Madagascar[1]
Moringa longituba Engl.	Kenya and Somalia[2]	Kenya, Ethiopia, and Somalia[1]
Moringa oleifera Lam.	India[2]	Various continents[1]
Moringa ovalifolia Dinter & Berger	Namibia, Angola[2]	Namibia and Angola[1]
Moringa peregrina (Forssk.) Fiori	Horn of Africa, Egypt[2]	Red Sea, Arabia, and Northeast Africa[1]
Moringa pygmaea Verdc.	Somalia[2]	North Somalia[1]
Moringa rivae Chiov.	Kenya and Ethiopia[2]	Kenya and Ethiopia[1]
Moringa ruspoliana Engl.	Ethiopia[2]	Kenya, Ethiopia, and Somalia[1]
Moringa stenopetala (Bak. f.) Cuf.	Kenya and Ethiopia[2]	Kenya, Ethiopia, and Somalia[1]

Sources: [1]Habtemariam, 2017; [2]Gandji et al., 2018.

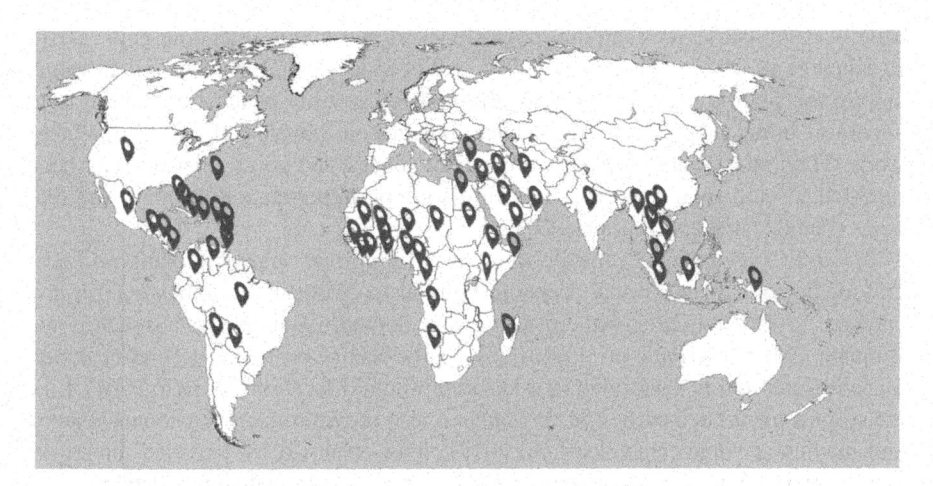

FIGURE 1.3 Geographical distributions of *Moringa* species.

medicine for centuries, using almost all parts of the plant to treat different diseases such as coughs, diarrhea, swelling, headaches, skin infections, and anemia, among other (Fahey, 2005; Padayachee and Baijnath, 2012; Ganguly, 2013; Leone et al., 2015). Its distribution and naturalization have spread to other tropical regions in more than 60 different countries worldwide (Dhakad et al., 2019). It has been reported in Southeast Asia, Western Asia, Africa, North and South America, on the islands of the Caribbean and West Indies; in Mexico, it is found on the Pacific coast from Baja California and Sonora (Velázquez-Zavala et al., 2016; Gandji et al., 2018).

The closest species to *M. oleifera* is *M. concanensis*, a medicinal plant that is present in the district of Perambalur, TamilNadu, India (Anbazhakan et al., 2007; Balamurugan and Balakrishnan, 2013; Santhi and Sengottuvel, 2016); it has been used for many years in communities in the Nilgiris region of India as an antifertility agent (Padayachee and Baijnath, 2012); it has also been used in traditional medicine to treat various diseases (paralysis, epilepsy, rheumatism, fainting, abscess, menstrual pain, constipation, jaundice, diabetes, skin tumors, headaches, dental problems, to reduce cholesterol levels and blood pressure, for the treatment of thyroid problems and leucorrhea and curing liver and spleen diseases) using the root and bark, the gum, the leaves, the flowers, and the fruits of the plant (Anbazhakan et al., 2007; Jayabharathi and Chitra, 2011; Padayachee and Baijnath, 2012).

The dry lands of Villamuthur, Perambalur, Veppan Thatthai, and Kunnam of Perambalur district are where *M. concanensis* is widely distributed (Manzoor et al., 2007). As for *M. peregrina*, it is one of the most economically important and valuable medicinal plants in the Egyptian desert, and could become one of the world's most valuable plants for its properties and economic importance; however, it is in danger of extinction due to overgrazing, and overexploitation for traditional medicines and other human activities (Boulos, 1999; Padayachee and Baijnath, 2012). The geographic distribution of this species is very wide; it can be found from the Dead Sea area, along the Red Sea to northern Somalia, and around the Arabian Peninsula to the mouth of the Arabian Gulf. The plant's occurrence is in Sistan-Baluchistan and Hormozgan provinces (Southeast Iran); *M. peregrina* usually inhabits crevices and the rocky slopes of mountains (Zaghloul et al., 2010; El-Kheir et al., 2017; Habtemariam, 2017). It is widely used for its medicinal value; among the plant's uses, it has been reported for treating headaches, fevers, constipation, burns, abdominal pains, back and muscle pains, and child-birth labor pains (Habtemariam, 2017). An infusion of leaves and roots in water has been reported to help treat malaria, stomach disorders, hypertension, asthma, and diabetes (Padayachee and Baijnath, 2012).

Moringa stenopetala, 'cabbage tree', is an important vegetable plant with several medicinal and nutritional properties similar to those found in *M. oleifera*; it is endemic to East Africa (Northern Kenya and Southern Ethiopia) (Padayachee and Baijnath, 2012; El-Kheir et al., 2017); usually occurs on rocky ground near permanent water and is also found in arid to semi-humid locations (Bosch, 2004). Like the species mentioned earlier, *M. stenopetala* also provides many nutritional benefits and medicinal value (Habtemariam, 2017); Rani et al. (2018) reported among its main traditional uses the treatment of flu, diabetes and associated disorders, malaria, stomach pain, diabetes, hypertension, expelling retained placenta, stomach pain, visceral leishmanisis, wound healing and the common cold; making use of the different parts of the plant, like the leaves and root. Nibret and Wink (2010) found that the extracts of *M. stenopetala* can be natural antimicrobial agents against bacteria that cause waterborne diseases. *Moringa oleifera*, *M. concanensis*, *M. stenopetala*, and *M. peregrina* are the four species among the 13 of the genus with the most research reported in the literature.

The other species (*M. arborea*, *M. borziana*, *M. drouhardii*, *M. hildebrandtii*, *M. longituba*, *M. ovalifolia*, *M. pygmaea*, *M. rivae*, and *M. ruspoliana*) are endemic to Madagascar (Boukandoul et al., 2018; Tahir et al., 2020). Some traditional uses have been reported for these species, mainly of medicinal value. *Moringa drouhardii* bark is used for colds and coughs; this species naturally inhabits scrubs on exposed rocky limestones or densely vegetated slopes (Padayachee and Baijnath, 2012). *Moringa rivae* has been reported to treat weakness of the thigh and calf muscles using the leaves of the plant; it is known to occur in the southern region of Ethiopia and also in Kenya (Rani et al., 2018; Habtemariam, 2017). *Moringa ruspoliana* is another example of the less-investigated *Moringa* species. It is distributed in Ethiopia and Somalia, and is used in the treatment of abdominal pains, eye and throat infections, sexually transmitted diseases in humans, tsetse fly bites, and livestock (Padayachee and Baijnath, 2012; Rani et al., 2018).

1.4 AGRICULTURAL RESEARCH

Agricultural research mainly focuses on the cultivation of *M. oleifera*, which is commonly cultivated throughout the world and has more information available in the literature. The uses of the *M. oleifera* tree in agriculture are well reported. *Moringa* plays an important role in food security, driving improved nutrition through nutrient-dense foods. The following uses in agriculture are highlighted.

Agroforestry, silviculture, or forestry: M. oleifera meets the requirements to be one of the agroforestry system components, due to its fast-growing, easily established, short rotation, deep root system, diverse uses, and food security potential (Kumar et al., 2017). *Moringa* trees are planted for multiple purposes, such as reducing soil erosion and acting as a windbreak; they are planted between crops interspersed with high-value others like vegetables increasing crop yield; where *Moringa* is integrated with agricultural produce it withdraws subsurface nutrients through its deep root system (Yasmeen et al., 2012; Palada et al., 2019).

Fertilizer: Moringa is considered a potential source of nutrients, which can be incorporated into the soil as green manure, that usually performs various functions such as improvement and protection of the soil, providing organic compounds and additional nutrients (Palada et al., 2019). Pod shucks and seed kernel press cake can be used as mulch and enhances soil fertility; the lower lignin and chitin content in the detritus matters of tree parts make it easily decomposable, and when they decompose they provide essential nutrients such as nitrogen (Mridha, 2015) for growth and development of plants; in this way, the productivity of the crops is increased by adding so many nutrients (Emmanuel et al., 2011; Kumar et al., 2017).

Erosion control, boundary, barrier, and support: The *Moringa* plants are used for ecosystem services such as erosion control; where strong winds and long, dry spells occur; as crop protection against the wind; they reduce sandstorms, being a barrier in crops; as well as support for climbing garden plants. They are also used as live fences and boundaries in farmlands (Mridha, 2015; Palada et al., 2019).

Biopesticide: Due to the overuse and misuse of pesticides and fungicides to manage pests and plant pathogens, harmful effects (death, cancer, allergies, congenital disorders, etc.) have been caused in humans and animals. These have led to the search for different pathogen control methods, focusing on non-synthetic chemicals for disease management in the agricultural sector (Abd El-Hack et al., 2018). It has been reported in various studies that *M. oleifera* is known to be one of the most effective plants as a natural biopesticide and inhibitor of several plant pathogens (Nwankwo et al., 2015; El-Masry et al., 2017). Compared to chemical methods, these bio-agents are environmentally friendly; thus, they can be included in integrated pest management strategies (Abd El-Hack et al., 2018).

Among other important uses that have been given to *M. oleifera* are *biogas*, an alternative biofuel, which can be produced from fresh *Moringa* biomass (Ahaotu et al., 2018; Palada et al., 2019); it has also been used as *animal and aquaculture feeds* due to its high content of protein, carotenoids, vitamins and minerals (mainly ascorbic acid and iron, respectively), and phytochemicals like kaempferitrin, isoquercitrin, rhamnetin, kaempferol, and quercetin (Mridha, 2015; Castillo et al., 2018; Pérez et al., 2019). The species has wide *ornamental* use in many different countries, as a garden plant, shade tree, and hedge plant (Abd El-Hack et al., 2018; Palada et al., 2019).

1.4.1 CULTIVATION AND PROPAGATION

Moringa oleifera is known as the 'never die' plant because of its large-scale adaptability to climate, soil, and other environmental variations (Boukandoul et al., 2018). It is classified as a tropical plant; however, it adapts to various agroclimatic conditions (Palada et al., 2019). This species grows best in a temperature range of 25–35°C but can tolerate up to 48°C or a light frost for a limited period. Moringa is drought-tolerant with a rainfall range of 250–3000 mm (the irrigation is needed for leaf production if rainfall < 800 mm); the best altitudes for *Moringa* are below 600 m, but it can also grow in altitudes up to 1200 m (Pandey et al., 2011; Leone et al., 2015; Mallenakuppe et al., 2019; Palada et al., 2019). Although it can grow in almost all soil types (except rigid and heavy clays) the plant prefers a well-drained sandy loam or sandy loam soil; it does not tolerate frequent flooding and poor drainage (Ramachandran et al., 1980; Boukandoul et al., 2018). This plant is tolerant of a wide soil pH range of 5–9, growing quite well in slightly acidic to highly alkaline pH (Mallenakuppe et al., 2019). It has been reported that the key factors that influence the growth and cultivation of *Moringa* are temperature, moisture, latitude, and elevation (Palada et al., 2019).

Moringa has bisexual flowers, highly pollinated, and pollination is mainly facilitated by animals as; bees and sunbirds. The seeds are also strongly winged, which may allow them to be spread short distances from the parent tree by wind, helping pollination (Gandji et al., 2018).

Moringa propagation develops in two main ways: sowing and cutting; direct sowing is the most adopted method due to its high germination rate (Leone et al., 2015), which ranges between 60% and 90% and occurs between 7 and 30 days after sowing (Parrota, 2005); another benefit is the formation of a deep, stout taproot with a wide-spreading system of thick, tuberous lateral roots (an advantage for stabilization and access to water) (Boukandoul et al., 2018). Sowing requires the selection of the seeds when they are easily available. It is used when the human labor is limited, it should be sown without pretreatment, as scarification does not facilitate germination, and the optimal sowing depth is 1–2 cm (Parrotta, 2005; Leone et al., 2015). When planting in a nursery is planned, the seedlings can be transplanted 3–6 weeks after germination when they reach around 30 cm (Gandji et al., 2018). However, propagation by cuttings is preferred by some farmers; it promises quick flowering and fruiting rates and gives the best quality fruits, although they are more sensitive to drought and wind because they have much shorter roots (Boukandoul et al., 2018). In the same way, cutting is preferred when seed availability is scarce or when labor is not a limiting factor. Usually, in Sudan, the seeds are preferred and in India, Indonesia, and in some areas of West Africa vegetative propagation is common (Leone et al., 2015).

The *M. oleifera* plant is fast-growing; it can reach 12 m up to the crown, and depending on cultivar, produces its first fruits between 6 and 12 months after planting. Trees can flower and fruit in one year and multiple seed harvests are possible in many parts of the world. Fruits (pods) ripen about 3 months after flowering and must be harvested as soon as possible. Each pod contains approximately 26 seeds of 1 cm diameter (Shindano and Kasase, 2009; Leone et al., 2015).

The plantation can be designed with spacing to produce seeds and leaves; seed production requires a low-density plantation (2.5 × 2.5 m or 3 × 3 m) in a triangular pattern. Leafs production can vary from intensive to semi-intensive or integrated into an agroforestry system, with the following spacings: intensive (10 × 10 cm to 20 × 20 cm) with a harvest interval between 35 and 45 days; semi-intensive (50 cm × 100 cm) with a harvest interval between 50 and 60 days; and an agroforestry system with a spacing distance of 2–4 m between rows, and a harvest interval around 60 days (Leone et al., 2015; Boukandoul et al., 2018; Gandji et al., 2018).

1.4.2 PESTS AND DISEASES

Usually, the *Moringa* tree is resistant to most serious pathogenic pests and diseases. However, it is susceptible to several insect pests, which can be categorized as borers (internal feeders), defoliators (leaf feeders), sucking insects (sap feeders), and non-insect pests (Kotikal and Math, 2016). These include the bark-eating caterpillar (*Indarbela quadrinotata*); the hairy caterpillar (*Eupterote mollifera*); the green leaf caterpillar (*Noorda blitealis*); *Tetragonia siva, Metanastia hyrtaca,* and *Heliothis armigera* caterpillars; the budworm (*Noorda moringae* Tams); an aphid (*Aphis caraccivera*); scale insects (*Ceroplastodes cajani* and *Diaspidotus* sp.); the stem borers (*Indarbela tetraonis* and *Diaxenopsis apomecynoides*); and a fruitfly (*Gitonia* sp.) (Ramachandran et al., 1980; Butani and Verma, 1981; Parrotta, 2005; Boukandoul et al., 2018).

The caterpillars feed, making a zigzag below the bark, and later bore inside the bark or main stem. The leaf-eating caterpillar is considered to be the most serious pest of annual *Moringa* as it occurs throughout the year and causes serious damage to the crop (Kotikal and Math, 2016); the leaves and seeds are affected by insect larvae such as *N. blitealis, Eupterote mollifera, Euproctis pasteopa, and Ulopeza phaeothoracica,* causing reduced leaf biomass and making these leaves improper for human consumption; seeds are punctured by larva and became unviable for seedling and other uses (Kotikal and Math, 2016; Palada et al., 2019). Some *Moringa* species have been reported to be more resistant or susceptible to attack by pests and diseases. *Moringa oleifera* is more susceptible to a fungal disease that specifically attacks its root system (Vaknin and Mishal, 2017), while *M. stenopetala* is more vulnerable to the seexxe caterpillar. Mite pests have been found to infect the leaves of species such as *M. rivae, M. concanensis,* and *M. oleifera* (Joshi et al., 2016; Boukandoul et al., 2018).

Although diseases and pests in *Moringa* are not considered a serious problem, it is important to follow agricultural measures and practices to avoid a decrease in production and economic losses.

1.5 CONCLUSIONS

Moringa is the only genus of the family Moringacea, classified taxonomically in the Capparales order. The genus *Moringa* comprises 13 species of tropical and subtropical flowering trees. *Moringa oleifera* is the most studied species; it is a plant considered multipurpose for all its applications. *Moringa* covers a very diverse range of growth habits or forms, from herbs and shrubs to large trees, which are usually resistant to most serious pathogenic pests and diseases. A set of unique traits distinguishes the Moringaceae family, among which large and pinnate leaves, divided into leaflets on a rachi, leaf glands borne from a short stem, conspicuous pedunculated glands where the petiole intersects the stem, fruits forming a long and woody capsule, and seeds with three longitudinal wings in most species with an outer mesotesta of thin-walled cells stand out. It is presumed that all *Moringa* species are native to India and Africa but have been distributed into several tropical countries over the years. Agricultural research is mainly focused on the cultivation of *M. oleifera,* for its main uses in agroforestry, silviculture, erosion control, as well as being a fertilizer, a boundary, a barrier, and a support. *Moringa oleifera* also has other important benefits for food, medicinal, and ornamental applications, water purification, biodiesel, biogas, and animal–aquaculture feeds production. The cultivation of *Moringa* is relatively easy; it can be propagated by both sexual and asexual means, with low demand for nutrients from the soil and water; the *Moringa* seed has almost all the essential nutrients in adequate quantities for its maintenance, rapid growth, and production. However, there are some limitations in the cultivation of *Moringa* that slow down the adoption of the species, so it is necessary to increase the level of use and cultivation of the species by identifying factors such as adequate planting densities, cutting frequencies, and information on agronomic practices to obtain higher biomass with good nutritional quality that will influence its adoption. This crop's perspective requires the genetic improvement of the species and the identification of industrial value ecotypes with a higher percentage of seed oil, proteins, bioactive compounds, and wide adaptability to the agroclimatic conditions to help in its adoption.

REFERENCES

Abd El-Hack, M.E., M. Alagawany, A.S. Elrys, E.S.M. Desoky, H.M.N. Tolba, A.S.M. Elnahal, et al. 2018. Effect of forage *Moringa oleifera* L. (Moringa) on animal health and nutrition and its beneficial applications in soil, plants and water purification. *Agriculture* 8: 33–54.

Abdel-Hameed, U.K. 2015. Molecular phylogenetics of Moringaceae martinov with emphasis on ethnomedicinal plant *Moringa oleifera* Lam. grown in Egypt. *Scholars Academic Journal of Biosciences* 3: 139–142.

Abe, R. and K. Ohtani. 2013. An ethnobotanical study of medicinal plants and traditional therapies on Batan Island, the Philippines. *Journal of Ethnopharmacology* 145: 554–565.

Ahaotu, E.O., R.E. Uwalaka, M.C. Edih and P.O. Ihiaha. 2018. Extraction of oil, biogas and biodiesel from *Moringa oleifera* seeds. *Annals of Agricultural Science, Moshtohor* 56: 1021–1026.

Anbazhakan, S., R. Dhandapani, P. Anandhakumar and S. Balu. 2007. Traditional medicinal knowledge on *Moringa concanensis* Nimmo of Perambalur District, Tamilnadu. *Ancient Science of Life* 26: 42–45.

Arora, D.S., J.G. Onsare and H. Kaur. 2013. Bioprospecting of *Moringa* (Moringaceae): microbiological perspective. *Journal of Pharmacognosy and Phytochemistry* 1: 193–215.

Azza, S.M. 2014. Morpho-anatomical variations of leaves and seeds among three *Moringa* species. *Life Science Journal* 11: 827–832.

Balamurugan, V. and V. Balakrishnan. 2013. Evaluation of phytochemical, pharmacognostical and antimicrobial activity from the bark of *Moringa concanensis* Nimmo. *International Journal of Current Microbiology and Applied Sciences* 2: 117–125.

Bosch, C.H. 2004. *Moringa stenopetala* (Baker f.) Cufod. In Grubben, G.J.H. and Denton, O.A. (eds.), *Plant Resources of Tropical Africa 2: Vegetables*, 395–399. Wageningen: PROTA Foundation.

Boukandoul, S., S. Casal and F. Zaidi. 2018. The potential of some moringa species for seed oil production. *Agriculture* 8: 150.

Boulos, L. 1999. Flora of Egypt. *Nordic Journal of Botany* 1: 417.

Butani, D.K. and S. Verma. 1981. Insect pests of vegetable and their control: drumsticks. *Pesticides* 15: 29–31.

Castillo, R.I., M.A. Angulo, R. Gutiérrez, M.D. Muy, J. León, J.B. Heredia. 2016. Epidemiological evidence suggesting the use of *Moringa oleifera* for decreasing cancer risk and processes related with its occurrence. *Journal of Pharmaceutical Biology* 6: 74–81.

Castillo, R.I., J. León, M.A. Angulo, R. Gutiérrez, M.D. Muy, J.B. Heredia. 2017. Nutritional and phenolic characterization of *Moringa oleifera* leaves grown in Sinaloa, México. *Pakistan Journal of Botany* 49: 161–168.

Castillo, R.I., J.J. Portillo, J. León, R. Gutiérrez, M.A. Angulo, M.D. Muy, and J.B. Heredia. 2018. Inclusion of moringa leaf powder (*Moringa oleifera*) in fodder for feeding Japanese quail (*Coturnix japonica*). *Brazilian Journal of Poultry Science* 20: 15–26.

Chhikara, N., A. Kaur, S. Mann, M.K. Garg, S.A. Sofi and A. Panghal. 2020. Bioactive compounds, associated health benefits and safety considerations of *Moringa oleifera* L.: an updated review. *Nutrition & Food Science* 51(2):255–277. doi:10.1108/NFS-03-2020-0087.

Dhakad, A.K., M. Ikram, S. Sharma, S. Khan, V.V. Pandey and A. Singh. 2019. Biological, nutritional, and therapeutic significance of *Moringa oleifera* Lam. *Phytotherapy Research* 33: 2870–2903.

El-Kheir, Z.A.A., A.M. Al-Zohairy, A.A. Dahab and S.R.A El-Fadl. 2017. Assessment of biodiversity based on morphological characteristics and rapid markers among three *Moringa* species. *Egyptian Journal of Biotechnology* 54: 45–61.

El-Masry, G.N., F.M. Saleh and S.A. Abd El-Mageed. 2017. Moringa plant, *Moringa oleifera* Lam., as a source of biopesticide. *Sinai Journal of Applied Sciences* 6: 285–292.

Emmanuel, S.A., S.G. Zaku, S.O. Adedirin, M. Tafida and S.A. Thomas. 2011. *Moringa oleifera* seed-cake, alternative biodegradable and biocompatibility organic fertilizer for modern farming. *Agriculture and Biology Journal of North America* 2: 1289–1292.

Fahey, J.W. 2005. *Moringa oleifera*: a review of the medical evidence for its nutritional, therapeutic, and prophylactic properties. Part 1. *Trees for Life Journal* 1: 5.

Gandji, K., F.J. Chadare, R. Idohou, V.K. Salako, A.E. Assogbadjo and R.G. Kakaï. 2018. Status and utilisation of *Moringa oleifera* Lam: q review. *African Crop Science Journal* 26: 137–156.

Ganguly, S. 2013. Indian ayurvedic and traditional medicinal implications of indigenously available plants, herbs and fruits: a review. *International Journal of Research in Ayurveda and Pharmacy* 4: 623–625.

Habtemariam, S. 2017. *The African and Arabian Moringa species. Chemistry, bioactivity and therapeutic applications.* Dordrecht: Elsevier.

Hassanein, A.M.A. and A.A. Al-Soqeer. 2018. Morphological and genetic diversity of *Moringa oleifera* and *Moringa peregrina* genotypes. *Horticulture, Environment and Biotechnology* 59: 251–261.

Hausiku, M.K., E.G. Kwembeya, P.M. Chimwamurombe and A. Mbangu. 2020. Assessment of species boundaries of the *Moringa ovalifolia* in Namibia using nuclear its DNA sequence data. *South African Journal of Botany* 131: 335–341.

Jaiswal, D., P.K. Rai, A. Kumar, S. Mehta and G. Watal. 2009. Effect of *Moringa oleifera* Lam. leaves aqueous extract therapy on hyperglycemic rats. *Journal of Ethnopharmacology* 123: 392–396.

Jayabharathi, M. and M. Chitra M. 2011. Evaluation of anti-inflammatory, analgesic and antipyretic activity of *Moringa concanensis* Nimmo. *Journal of Chemical and Pharmaceutical Research* 3: 802–806.

Joshi, R.C., B.V. David, R. Kant. 2016. A review of the insect and mite pests of *Moringa oleifera* Lam. *Agricultural Development* 29: 29–33.

Kasolo, J., G. Bimenya, L. Ojok, J. Ochieng and J. Ogwal-Okeng. 2010. Phytochemicals and uses of *Moringa oleifera* leaves in Ugandan rural communities. *Journal of Medicinal Plants Research* 4: 753–757.

Kotikal, Y.K. and M. Math. 2016. Insect and non insect pests associated with drumstick, *Moringa oleifera* (Lamk.). *Entomology, Ornithology & Herpetology: Current Research* 5: 180.

Kumar, Y., T.K. Thakur, M.L. Sahu and A. Thakur. 2017. A multifunctional wonder tree: *Moringa oleifera* Lam open new dimensions in field of agroforestry in India. *International Journal of Current Microbiology and Applied Sciences* 6: 229–235.

Lalas, S. and J. Tsaknis. 2002. Characterization of *Moringa oleifera* seed oil variety "Periyakulam 1". *Journal of Food Composition and Analysis* 15: 65–78.

León-López, L., Y. Escobar, J. Milán, D.M. Domínguez, R. Gutiérrez, and E.O. Cuevas. 2020. Chemical proximate composition, antinutritional factors content, and antioxidant capacity of anatomical seed fractions of *Moringa oleifera*. *Acta Universitaria* 30: e2892.

Leone, A., A. Spada, A. Battezzati, A. Schiraldi, J. Aristil and S. Bertoli. 2015. Cultivation, genetic, ethnopharmacology, phytochemistry and pharmacology of *Moringa oleifera* leaves: an overview. *International Journal of Molecular Sciences* 16: 12791–12835.

Mallenakuppe, R., H. Homabalegowda, M.D. Gouri, P.S. Basavaraju and U.B. Chandrashekharaiah. 2019. History, taxonomy and propagation of *Moringa oleifera* – a review. *SSR Institute of International Journal of Life Sciences* 5: 2322–2327.

Manzoor, M., F. Anwar, T. Iqbal and M.I. Bhanger. 2007. Physico-chemical characterization of *Moringa concanensis* seeds and seed oil. *Journal of the American Oil Chemists' Society* 84: 413–419.

Mridha, M.A.U. 2015. Prospects of moringa cultivation in Saudi Arabia. *Journal of Applied Environmental and Biological Sciences* 5: 39–46.

Nibret, E. and M. Wink. 2010. Trypanocidal and antileukaemic effects of the essential oils of *Hagenia abyssinica*, *Leonotis ocymifolia*, *Moringa stenopetala*, and their main individual constituents. *Phytomedicine* 17: 911–920.

Nwankwo, E.N., N.J. Okonkwo, C.U. Ogbonna, C.J. Akpom, C.M. Egbuche and B.C. Ukonze. 2015. *Moringa oleifera* and *Annona muricata* seed oil extracts as biopesticides against the second and fourth larval instar of *Aedes aegypti* L. *Journal of Biopesticides* 8: 56–61.

Olson, M.E. 2002a. Combining data from DNA sequences and morphology for a phylogeny of Moringaceae. *Systematic Botany* 27: 55–73.

Olson, M.E. 2002b. Intergeneric relationships within the Caricaceae–Moringacecae clade (Brassicales), and potential morphological synapomorphies of the clade and its families. *International Journal of Plant Sciences* 163: 51–65.

Olson, M.E. and J.W. Fahey. 2011. *Moringa oleifera*: a multipurpose tree for the dry tropics. *Revista Mexicana De Biodiversidad* 82: 1071–1082.

Olson, M.E. and S. Carlquist. 2001. Stem and root anatomical correlations with life form diversity, ecology, and systematics in *Moringa* (Moringaceae). *Botanical Journal of the Linnean Society* 135: 315–348.

Olson, M.E. and S.G. Razafimandimbison. 2000. *Moringa hildebrandtii* (Moringaceae): a tree extinct in the wild but preserved by indigenous horticultural practices in Madagascar. *Adansonia* 22: 217–221.

Osman, H.E. and A.A. Abohassan. 2012. Morphological and analytical characterization of *Moringa peregrina* populations in Western Saudi Arabia. *International Scientific Journal Theoretical & Applied Science* 4: 174–184.

Özcan, M.M. 2020. *Moringa* spp: composition and bioactive properties. *South African Journal of Botany* 129: 25–31.

Padayachee, B. and H. Baijnath. 2012. An overview of the medicinal importance of Moringaceae. *Journal of Medicinal Plants Research* 6: 5831–5839.

Palada, M.C., A.W. Ebert and R.C. Joshi. 2019. *The miracle tree:* Moringa oleifera. Estados Unidos: Xlibris.

Paliwal, R., V. Sharma and J. Pracheta. 2011. A review on horseradish tree (*Moringa oleifera*): a multipurpose tree with high economic and commercial importance. *Asian Journal of Biotechnology* 3: 317–328.

Pandey, A., K. Pradheep, R. Gupta, E.R. Nayar and D.C. Bhandari. 2011. 'Drumstick tree' (*Moringa oleifera* Lam.): a multipurpose potential species in India. *Genetic Resources and Crop Evolution* 58: 456–460.

Parrota, J. 2005. *Moringa. Agroforestería en las américas* 43: 54–61.

Pérez, K.J., R. Gutiérrez, A. Téllez, C.A. Gómez, C. Reyes, R.O. Navarro, et al. 2019. Effect of extrusion in the elaboration of an animal feed based on *Moringa oleifera* Lam and *Zea mays* as a partial substitute of fishmeal in the diet of the adult stage of *Oreochromis niloticus*. *Acta Universitaria* 29: e2266.

Ramachandran, C., K.V. Peter and P.K. Gopalkrishna. 1980. Drumstick (*Moringa oleifera*): a multipurpose Indian vegetable. *Economic Botany* 34: 276–283.

Rani, N.Z., K. Husain and E. Kumolosasi. 2018. *Moringa* genus: a review of phytochemistry and pharmacology. *Frontiers in Pharmacology* 9: 108.

Rodman, J.E., K.G. Karol, R.A. Price and K.J. Sytsma. 1996. Molecules, morphology, and Dahlgren's expanded order Capparales. *Systematic Botany* 21: 289–307.

Ruiz, A.I., M.I. Mercado, M.E. Guantay and G.I. Ponessa. 2019. Anatomía e histoquímica foliar y caulinar de *Moringa oleifera* (Moringaceae). *Boletín de la Sociedad Argentina de Botánica* 54: 325–343.

Santhi, K. and R. Sengottuvel. 2016. The antioxidant activity of the methanolic leaf extract of *Moringa concanensis* Nimmo. *International Journal of Advanced Multidisciplinary Research* 3: 1–5.

Shindano, J. and C. Kasase. 2009. Moringa (*Moringa oleifera*): a source of food and nutrition, medicine and industrial products. In *African Natural Plant Products: New Discoveries and Challenges in Chemistry and Quality*, ACS Symposium Series, 421–467. Washington, DC: American Chemical Society.

Sivasankari, B., M. Anandharaj and P. Gunasekaran. 2014. An ethnobotanical study of indigenous knowledge on medicinal plants used by the village peoples of Thoppampatti, Dindigul district, Tamilnadu, India. *Journal of Ethnopharmacology* 153: 408–423.

Spandana, U., P. Srikanth, J. Gopichand and V. Ashok Babu. 2016. A review on miracle tree: *Moringa oleifera*. *Journal of Pharmacognosy and Phytochemistry* 5: 189–191.

Sudha, P., S.M. Asdaq, S.S. Dhamingi and G.K. Chandrakala. 2010. Immunomodulatory activity of methanolic leaf extract of *Moringa oleifera* in animals. *Indian Journal of Physiology and Pharmacology* 54: 133–140.

Tahir, N.A., H.O. Majeed, H.A. Azeez, D.A. Omer, J.M. Faraj and W.R.M. Palani. 2020. Allelopathic plants: 27. *Moringa* species. *Allelopathy Journal* 50: 35–46.

Vaknin, Y. and A. Mishal. 2017. The potential of tropical "miracle tree" *Moringa oleifera* and its desert relative *Moringa peregrina* as edible seed-oil and protein crops under Mediterranean conditions. *Scientia Horticulturae* 225: 431–437.

Velázquez-Zavala, M., I.E. Peón-Escalante, R. Zepeda-Bautista and M.A. Jiménez-Arellanes. 2016. Moringa (*Moringa oleifera* Lam.): potential uses in agriculture, industry and medicine. *Revista Chapingo Serie Horticultura* 22: 95–116.

Verdcourt, B. 1985. A synopsis of Moringaceae. *Kew Bulletin* 40: 1–23.

Verdcourt, B. 1986. *Flora of Tropical East Africa –Moringaceae*. Boca Raton, FL: CRC Press/ Taylor & Francis.

Yabesh, J.E., S. Prabhu and S. Vijayakumar. 2014. An ethnobotanical study of medicinal plants used by traditional healers in silent valley of Kerala, India. *Journal of Ethnopharmacology* 154: 774–789.

Yasmeen A.Y., S.M.A. Basra and A. Wahid. 2012. Performance of late sown wheat in response to foliar application of *Moringa oleifera* Lam. leaf extract. *Chilean Journal of Agricultural Research* 72: 92–97.

Zaghloul, M.S., R.H. Abd El-Wahab and A.A. Moustafa. 2010. Ecological assessment and phenotypic variation of Sinai's remnant populations of *Moringa peregrina*. *Applied Ecology and Environmental Research* 8: 351–366.

2 Genetic Diversity of the *Moringa* Genus

Liseth Daniela Jaimes-Rangel
Universidad Francisco de Paula Santander, Ingeniería
Biotecnológica, San José de Cúcuta, Norte de Santander,
Colombia.

Diana Marcela Arias-Moreno
Grupo de Investigación BIOPLASMA, Facultad de Ciencias,
Universidad Pedagógica y Tecnológica de Colombia, Tunja,
Colombia.

Abraham Cruz-Mendívil
Cátedras CONACYT-Instituto Politécnico Nacional, CIIDIR
Unidad Sinaloa, Departamento de Biotecnología Agrícola,
Guasave, Sinaloa, México.
Corresponding author: acruzm@conacyt.mx

CONTENTS

DOI: 10.1201/9781003108863-2

2.1 INTRODUCTION

Genetic diversity, the fundamental source of biodiversity, is the measure that quantifies the variation found within a population of a given species. Genetic diversity between individuals reflects the presence of different alleles in the gene pool and, therefore, of different genotypes within populations. Information about plant genetic diversity and population structure is necessary to develop appropriate conservation biology strategies and assist plant breeding in selecting parents for crossing, providing a more rational basis for expanding the gene pool and identifying materials that harbor favorable alleles. Molecular phylogenetics and genetic diversity analyses can also help clarify the taxonomic identity and evolutionary relationships of the wild relatives of crop species (Poczai et al., 2012). The success of any plant breeding program mainly depends on the extent and magnitude of the genetic variability existing in the population. Hence, comprehensive and adequate knowledge of the genetics of various characters is essential for the improvement of both qualitative and quantitative traits (Tripathy and Mallikarjunarao, 2020).

Moringaceae are old-world perennial softwood trees indigenous to the western and sub-Himalayan tracts and are distributed in tropical regions of the world (Shahzad et al., 2013). Its only genus, *Moringa*, comprises 13 species and is remarkable for the great diversity of tree habits and floral morphology (Olson, 2002). The *Moringa* genus is well-known for its multiple uses: the seeds are used for purifying water, the leaves as nutrition supplements, the oil as a biofuel, the trunks as gum, the flowers as honey, and the whole plant can be used for medicinal purposes (Abd et al., 2018).

The 13 *Moringa* species can be categorized into three groups depending on their type of trunk. *M. stenopetala*, *M. drouhardii*, *M. ovalifolia*, and *M. hildebrandtii* have bloated water-storing trunks known as bottle trees. Meanwhile, *M. peregrina*, *M. concanensis*, *M. oleifera*, *M. arborea*, *M. rivae*, and *M. ruspoliana* have slender trunks. Finally, *M. borziana*, *M. pygmaea*, and *M. longituba* are tuberous shrubs endemic to Northeast Africa. Only *M. oleifera* has been subjected to breeding and is currently cultivated throughout the tropics worldwide, whereas the other 12 species have mainly local uses, and their distribution is more limited (Olson, 2002). The main objective of this chapter was to review the different molecular approaches used to analyze the genetic diversity of the *Moringa* genus and to suggest future applications in the field.

2.2 MOLECULAR MARKERS

Genetic markers occupy specific genomic positions within chromosomes (*loci*) and represent genetic differences between individual organisms or species. Markers close (linked) to genes often do not affect the phenotype of the trait of interest. However, these markers act as 'flags' that allow the trait to be traced in different populations and their offspring. There are three main types of genetic markers: (1) morphological markers that are themselves phenotypic traits or characters; (2) biochemical markers, which include allelic variants of enzymes called isoenzymes; and (3) DNA (or molecular) markers, which reveal sites of variation in the DNA sequence (Collard et al., 2005).

A molecular marker is a DNA sequence with a known location on the chromosome and whose inheritance can be monitored (Mosa et al., 2019). Molecular markers are extensively utilized to assess genetic diversity, infer systematics and molecular phylogeny, identify close relatives and varieties, manage and improve crops, and for conservation of endangered species, among other applications. Molecular markers may be dominant (unable to distinguish heterozygous individuals) or codominant (the allelic difference can be detected; Adhikari et al., 2017).

Molecular markers have several advantages in plant breeding because they are not affected by environmental conditions, are abundant in the genome, are more objective than phenotypic markers, and can be evaluated in seeds or early plant growth stages. Therefore, molecular markers can be used for marker-assisted selection (MAS) to accelerate the plant breeding programs compared to conventional methods (Kim et al., 2016). Resolution and usability of DNA marker systems improved considerably with advancements in molecular biology (Grover and Sharma, 2014). Depending on the method of analysis, DNA marker techniques can be classified into three categories: (1) based on hybridization, (2) based on polymerase chain reaction (PCR), and (3) based on DNA sequencing (Garrido-Cardenas et al., 2017). Next, the basics of the most popular molecular markers are described.

2.2.1 RESTRICTION FRAGMENT LENGTH POLYMORPHISM (RFLP)

RFLP was the first molecular marker technique and the only system based on hybridization (Nadeem et al., 2018). After agarose gel electrophoresis and Southern blotting, restricted fragments are hybridized to short, single-copy genomic or cDNA radioactively labeled probes and visualized by autoradiography. Different probes are screened in combination with several endonucleases, and each combination generates a unique RFLP pattern at a specific locus (Grover and Sharma, 2014). The principle of RFLPs is that base-pair deletions, mutations, inversions, translocations, and transpositions lead to the gain or loss of recognition sites for restriction enzymes, resulting in fragments of various lengths and polymorphism (Nadeem et al., 2018). Because point mutations may occur in each pair's chromosome, RFLP markers are designed to detect both alleles in a heterozygous sample (Garrido-Cardenas et al., 2017).

2.2.2 RANDOMLY AMPLIFIED POLYMORPHIC DNA (RAPD)

This technique is based on the amplification of genomic DNA by PCR using a single 10-base primer with random sequence and at least 60% GC content (Grover and Sharma, 2014). During PCR, amplification occurs when two hybridization sites are similar to each other and in opposite directions within approximately 2000 bases, obtaining fragments of different sizes (Nadeem et al., 2018). If the primer binding sites change due to a mutation, the amplification products will also change, obtaining a substantially different banding profile. The main advantage of RAPD is that it is not necessary to know the target DNA sequence, and one of the disadvantages is that most RAPD markers are dominant, and thus could not identify heterozygotes (Garrido-Cardenas et al., 2017).

2.2.3 AMPLIFIED FRAGMENT LENGTH POLYMORPHISM (AFLP)

The limitations of the RAPD and RFLP markers were overcome by the development of AFLP markers, which combine DNA digestion and PCR amplification. In AFLP, two restriction enzymes (a frequent cutter and a rare cutter) are used to digest genomic DNA. Each fragment end is ligated to primers containing a 3-bp recognition site for the rare cutter and a 6-bp recognition site for the frequent cutter (Nadeem et al., 2018). The primers also carry random extensions of 1–3 bases at the 3' end to selectively amplify genomic fragments cut by both enzymes at random genome locations (Grover and Sharma, 2014). An advantage of AFLP markers is that they can be easily multiplexed, which greatly increases their performance. The main disadvantage is that fragments with low sequence homology between samples will produce a very low number of common AFLPs (Garrido-Cardenas et al., 2017).

2.2.4 MICROSATELLITES

Microsatellites, also known as simple sequence repeats (SSR), represent a class of repetitive sequences widely distributed in all genomes. They consist of arrays of tandemly repeated short nucleotide motifs of 1–6 bases, which tend to be imprecisely replicated during DNA synthesis, generating new alleles with different numbers of repeating units (Grover and Sharma, 2014). SSR are present in both coding and noncoding regions of all genomes and may also be present in chloroplast and mitochondrial DNA (Garrido-Cardenas et al., 2017). The sequences flanking the SSRs are mostly conserved and used in the development of primers for PCR amplification. SSR markers are considered a marker of choice, as they are codominant, with high reproducibility and greater genome abundance, and they can be used efficiently in plant mapping studies (Nadeem et al., 2018).

2.2.5 INTER SIMPLE SEQUENCE REPEAT (ISSR)

This technique is based on the amplification of DNA segments located between two regions of identical microsatellites but of opposite orientation. In general, primers for ISSR are 15–30 bases in length with 1–4 degenerate bases at the 3' or 5' end, which are extended into the flanking sequences. ISSRs are simple and easy to use and do not need prior knowledge of DNA sequences. However, they are dominant markers and have low reproducibility (Nadeem et al., 2018).

2.2.6 RANDOMLY AMPLIFIED MICROSATELLITE POLYMORPHISM (RAMP)

RAMP markers were developed to overcome the disadvantages of SSR and RAPD. This marker system involves an SSR primer utilized to amplify genomic DNA in the absence or presence of RAPD primers. RAMP markers are cost-effective, reflect higher polymorphism, have a wide distribution in the genome, and have been successfully applied in various plants for molecular characterization (Nadeem et al., 2018).

2.2.7 SEQUENCE-RELATED AMPLIFIED POLYMORPHISM (SRAP)

This marker system is based on the amplification of open reading frames (ORFs) using two primers of 17–18 bases long, containing the CCGG sequence in the forward primer and AATT in the reverse primer. SRAPs are dominant markers in nature, and DNA fragments are scored by the presence or absence of a band (Nadeem et al., 2018). SRAP markers have been used primarily to develop quantitative trait loci (QTL) in advanced hybrids and assess large germplasm collections' genetic diversity. However, they also can be used in plant systematics biogeography, conservation, and ecology (Robarts and Wolfe, 2014).

2.2.8 INTERNAL TRANSCRIBED SPACER (ITS)

DNA barcoding is an approach to identify species based on sequences from a short, standardized DNA region. The ITS and ITS2 regions of nuclear ribosomal DNA possess valuable characteristics as DNA barcodes, such as the availability of conserved regions for designing universal primers, the ease of its amplification, and sufficient variability to distinguish even closely related species within different families and genera (Yao et al., 2010).

2.2.9 START CODON TARGETED POLYMORPHISM (SCoT)

Dominant markers that are based on the short-conserved region flanking the ATG start codon in plant genes. This method uses a single 18-mer primer in PCR and an annealing temperature of 50°C (Collard and Mackill, 2009). SCoT markers have been used successfully to assess genetic diversity and structure, identify cultivars, QTL mapping, and DNA fingerprinting in different plant species (Etminan et al., 2016).

2.2.10 CYTOCHROME P$_{450}$ (CYTP450)

In higher plants, CytP450 monooxygenases play important roles in oxidative detoxification and the biosynthesis of secondary metabolites. Intra- and interspecific variations in CytP450 sequences of 51 plant species from 28 taxonomic families reflected the diversity of functional regions in the plant genome, suggesting that CytP450 would be effective genetic markers for the assessment of genetic diversity in plants, including non-model species (Yamanaka et al., 2003).

2.2.11 SINGLE NUCLEOTIDE POLYMORPHISM (SNPs)

Single base-pair changes present in the genome sequence of an individual are known as SNPs and might be the result of transitions (C/T or G/A), transversions (C/G, A/T, C/A, or T/G), or insertion/deletions (InDels; Nadeem et al., 2018). SNPs allow evaluation of a large number of loci, effectively discriminating between homozygous and heterozygous alleles. Furthermore, SNPs are homogeneously distributed throughout the genome, have low mutation rates, and show high heritability, making them ideal markers (Garrido-Cardenas et al., 2017). Recent advances in microarray and next-generation sequencing (NGS) technologies have made it possible to develop high-throughput genotyping methods that simultaneously allow the discovery and assay of thousands of SNPs.

2.3 GENETIC DIVERSITY OF THE *MORINGA* GENUS USING MOLECULAR MARKERS

In the *Moringa* genus, the uses of molecular markers have been focused on assessing genetic diversity and population structure. In recent years, several DNA-based molecular markers have been developed to provide detailed information about the diversity patterns within natural populations and germplasm collections of *Moringa*. Additionally, molecular markers have been used to detect differences at the molecular loci between closely related individuals to select suitable breeding lines in *M. oleifera* (Kumar et al., 2017). The studies of genetic diversity using molecular markers have been important for designing strategies for the conservation and exploitation of *Moringa* genetic resources. However, there is still limited knowledge of the available genetic diversity in the *Moringa* genus (Gandji et al., 2018). Recently, some efforts have been focused on the association of phenotypic traits variability with the presence of molecular markers (Rajalakshmi et al., 2019; Panwar and Mathur, 2020). The previous studies were mainly concentrated on *M. oleifera*; however, other *Moringa* species are gaining more attention due to nutritional, industrial, and medicinal uses (Senthilkumar et al., 2018). Next, we will describe the application of molecular markers in the study of the genetic diversity of the *Moringa* genus.

2.3.1 *MORINGA OLEIFERA*

Commonly known as the miracle tree, *M. oleifera* is the most widely cultivated and well-known species in the monogeneric family Moringaceae (Olson, 2002). The biological properties of *M. oleifera* have been studied since the 1970s, and *Moringa* has been named the most nutrient-rich plant (Abd et al., 2018). This species originates from northern India and is currently distributed throughout almost all tropical countries, increasing the species' variability. However, the number of accessions in collections and active germplasm banks is incipient worldwide (Leone et al., 2015). Several studies have been conducted with dominant and codominant markers to research the genetic diversity of *M. oleifera*, which are described below.

The first study of genetic diversity in *M. oleifera* was conducted by Muluvi et al. (1999) using AFLP markers to analyze 140 individuals from seven populations: two from India, one from Malawi, and four from Kenya. Four pairs of AFLP primers generated 236 bands, of which 157 (66.5%) were polymorphic. The analysis of molecular variance (AMOVA) showed that 59.1% of the genetic variability existed among populations, with a significant variation (14.4%) between the Kenyan population and the others. A phenetic tree based on genetic distance values showed a clear separation of the two Indian populations from the four Kenyan populations, suggesting two sources of germplasm introductions to Kenya. Meanwhile, the Malawi individuals were dispersed within the two Indian populations, supporting the hypothesis that the Malawi population is a recent introduction from India.

On the other hand, Mgendi, Manoko, and Nyomora (2010) used 12 RAPD markers to analyze the genetic diversity in four cultivated and three non-cultivated populations of *M. oleifera* from coastal regions of Tanzania. Cluster analysis showed five groups with a separation between cultivated and non-cultivated *M. oleifera* populations, attributed to genetic changes due to selection pressure that allows fixation of a particular character. However, the clustering was not according to the geographical areas where the collection was conducted, suggesting high gene flow rates between populations. Furthermore, Cruz da Silva et al. (2012) used 17 RAPD markers to characterize the genetic diversity of 16 *M. oleifera* accessions from the Germplasm Bank of Embrapa (Brazil). A total of 95 fragments were generated, of which 59 were polymorphic. Principal coordinates analysis (PCoA) allowed the identification of four groups. The mean values of polymorphism information content (PIC) and genetic diversity index (H) were 0.22 in both cases. These results indicate a low genetic diversity, suggesting the need to integrate and characterize new accessions to the Embrapa Germplasm Bank. Similar results were reported by Rufai et al. (2013), who evaluated the genetic diversity of 20 *M. oleifera* genotypes from Malaysian, USA, Thailand, India, Tanzania, and Taiwan, using 20 RAPD markers. The mean percentage of polymorphism was 52.73%, and the mean expected heterozygosity (H_e) was 0.156. The AMOVA revealed that 95% of the genetic variation occurred within the population, consistent with the low genetic differentiation observed among the populations ($F_{st} = 0.16$). On the other hand, Popoola et al. (2014) reported that the clustering pattern at both morpho-agronomic and molecular levels did not match the geographical distribution of 13 *M. oleifera* accessions from Nigeria using five RAPD markers.

Saini et al. (2013) combined 25 RAPD, 10 ISSR, and 11 CytP450 markers to evaluate the genetic variability in eight commercial cultivars of *M. oleifera* collected from different Indian states. The comparative analysis among marker systems showed that ISSR was better for assessing genetic diversity in *M. oleifera*, because it showed the highest marker index (MI). The cluster analysis based on the three sets of marker data indicated that cultivars were not clustered according to their geographical origin. This suggests the spread of planting material and high rates of gene flow through cross-pollination.

Hassan et al. (2020) used 10 SCoT and 10 ISSR markers to assess the genetic diversity of 10 *M. oleifera* accessions from Egypt. Additionally, DNA barcoding (*rbcL, matK, psbA,* and *ITS* genes) was used to understand the genetic relationships among the *M. oleifera* accessions. Low levels of polymorphism were observed with ISSR and SCoT markers producing 19.7% and 19.4% of polymorphic bands, respectively. A phylogenetic neighbor-joining (NJ) tree was constructed based on DNA barcodes and showed that *M. oleifera* accessions were clustered into two clades with 99% similarity. These results suggest that DNA barcoding could be a useful tool for genetic similarity detection in *M. oleifera* accessions.

Twenty polymorphic SSR markers were developed to determine population genetic parameters in 24 domesticated individuals of *M. oleifera* from India and Myanmar (Wu et al., 2010). The mean values of expected (H_e) and observed (H_o) heterozygosity were 0.5455 and 0.4562, respectively. Significant departures from Hardy–Weinberg equilibrium were detected in all loci, suggesting inbreeding pressure and limited sample size. Also, Shahzad et al. (2013) used SSR markers to study the genetic diversity among 161 *M. oleifera* accessions, from which 131 were collected in Pakistan and 30 were obtained from the ECHO (Educational Concerns for Hunger Organization) collection. The 19 SSR markers proved highly inform-ative, with a mean PIC of 0.59 and a mean H_o of 0.58. In general, a higher gen-etic diversity was observed in wild accessions from Pakistan. In contrast, a lower genetic diversity was observed in cultivated accessions from ECHO, supporting the Indo-Pakistan origin of *M. oleifera* and emphasizing the great potential of these accessions for plant breeding programs.

Furthermore, Ganesan et al. (2014) used 19 SSR markers to analyze the genetic diversity of 300 genotypes from 12 natural populations of *M. oleifera* collected in India. The AMOVA revealed that 95% of the variation was found within populations, 3% among populations, and 2% among regions (northern and southern India). No clear geographical isolation was observed in the populations, suggesting that individ-uals from different geographical areas are not genetically distinct. On the other hand, the population structure analysis grouped the accessions into five populations. The 23 genotypes from northern India were distributed only in two clusters, supporting the hypothesis that *M. oleifera* originated in the north of India and then moved to southern India and got further diversified. Finally, Amao et al. (2017) evaluated the genetic diversity of 31 *M. oleifera* accessions from Nigeria using 10 SSR markers. The mean values of PIC and gene diversity were 0.4813 and 0.5224, respectively. Cluster ana-lysis grouped the 31 accessions into two main clusters with four subclusters. Six accessions were duplicated or closely related, and some accessions were grouped based on the area of collection. However, there was no clear geographical isolation of the accessions studied.

On the other hand, Chaves-Bedoya et al. (2017) used six RAMP markers to ana-lyze the genetic diversity of 45 *M. oleifera* accessions from four regions of Colombia. RAM markers produced a total of 140 bands, of which 74% were polymorphic. The mean H_e was 0.18, suggesting a low genetic diversity in Colombian accessions, which could be derived from a common population. Cluster analysis grouped the 45 accessions into four clusters; the first comprised 32 accessions and some clustering patterns were observed according to their geographical origin.

The selection of suitable breeding lines is crucial for the development and release of new varieties. In this sense, Kumar et al. (2017) used a combination of 31 RAPD, 10 CytP450, and seed protein markers to characterize the genetic variation in seven advanced breeding lines of *M. oleifera* from India. CytP450 markers were more effi-cient than RAPDs for variability analysis, as evidenced by a higher MI. Cluster ana-lysis based on both marker data grouped the seven advanced breeding lines into two main clusters. The first cluster was composed only by the cv. Dhanraj, suggesting that the other six lines are derived from a single mother source.

On the other hand, the seed protein profiles could only clearly differentiate the G5 and G1 lines from the other lines. Therefore, variation among other breeding lines could not be established using this marker system. Similarly, a genetic diversity study was carried out in superior genotypes of *M. oleifera* collected from India's different states using CytP450 markers (Ravi et al., 2020). A yield trial was conducted in 120 *M. oleifera* trees, and 23 genotypes with fruit yield 50% higher than the average were selected for genetic diversity analysis. CytP450 markers showed a high discriminatory potential to distinguish among the genotypes evaluated. According to their geographical origin, the accessions were not clustered, which may be due to the out-crossed pollination of *M. oleifera* that is expected to maintain more variation within populations than between populations. Moreover, the high genetic diversity observed in these superior genotypes with high fruit yield could be exploited in breeding programs to obtain new cultivars.

Genetic and environmental factors influence the accumulation of nutritional and nutraceutical compounds in plants, and its variation can be correlated with the presence of molecular markers. For instance, Rajalakshmi et al. (2019) utilized 39 SRAP and 24 ISSR markers to analyze the genetic diversity of 97 *M. oleifera* accessions from India, as well as to associate these markers with leaf iron content. Both markers showed similar results regarding the estimates of diversity parameters, but ISSRs were slightly more informative than SRAPs. The AMOVA showed that 86% of the genetic variation was observed within populations, and 14% among populations, which was in agreement with the low genetic differentiation observed among populations (G_{ST} = 0.15). The structure, cluster, and PCoA analyses divided the 97 genotypes into two groups. The first comprised Tamil Nadu's accessions, whereas the second included Andhra Pradesh and Odisha.

Interestingly, the regression analysis identified two markers (SRAP4F3R and ISSR26) associated with leaf iron content, which could be used in MAS programs. In a similar approach, Panwar and Mathur (2020) analyzed the genetic diversity of 57 *M. oleifera* accessions from India using 20 RAPD markers. Additionally, their content of phenolic compounds was assayed using high-performance liquid chromatography (HPLC). In this study, a possible association between the genetic and phytochemical diversity of *M. oleifera* was observed because the accessions grouped in cluster-I showed higher concentrations of phenolics compounds. Thus, these accessions could be used for breeding new varieties with high nutraceutical value.

2.3.2 *MORINGA PEREGRINA*

Moringa peregrina is the most drought-tolerant species of the *Moringa* genus, it is native to semi-arid and arid areas from Arabian Peninsula, Northeast Africa, and Southwest Asia, and has a wide range of uses, including traditional, nutritional, industrial, and medicinal (Senthilkumar et al., 2018). However, its wild populations are declining due to overexploitation, habitat destruction by human activities, and low seedling recruitment (Dadamouny et al., 2016). Moreover, the conservation status of *M. peregrina* is unknown due to the limited research carried out to determine its population size and structure, as well as its occurrence and occupancy areas (Robiansyah et al,. 2014). Thus, conservation strategies are required to maintain the valuable genetic resources of *M. peregrina*.

In this regard, Al Khateeb et al. (2013) established a micropropagation protocol for *M. peregrina*, and the genetic stability of micropropagated plants was assessed using 10 ISSR markers. Identical banding patterns were observed among the mother plant and the micropropagated shoots after eight subcultures. These results suggested that shoot proliferation could be used for clonal propagation of *M. peregrina* with minimal risk of producing somaclonal variants. Furthermore, Alaklabi (2015) used an ITS marker to analyze the genetic stability of 14 *M. peregrina* accessions from four regions of Saudi Arabia. A fragment of 344 pb of the *ITS2* gene was amplified in all the *M. peregrina* accessions and failed to amplify in other species, suggesting that *ITS2* is a reliable marker to identify individuals of *M. peregrina* that remain undifferentiated morphologically. Finally, Hassanein and Al-Soqeer (2018) evaluated the diversity of *M. oleifera* and *M. peregrina* genotypes from Saudi Arabia using 14 morphological traits and 10 ISSR markers. Plant height and pinnate dimensions were the most pertinent phenotypic traits for the discrimination among genotypes. ISSR markers showed a high level of polymorphism (90.8%) which could be sufficient for characterizing the genetic diversity in *Moringa*. The PCoA from molecular data separated the two species into two main clusters, and a higher variance was observed between *M. oleifera* genotypes. Interestingly, two ISSR markers showed a high correlation with some morphological traits; therefore, they could be very useful to identify *Moringa* species.

2.3.3 MORINGA STENOPETALA

Commonly known as 'cabbage tree', *M. stenopetala* was domesticated in the east African lowlands and is native to southern Ethiopia. Many different ecotypes are found, and varieties are cultivated as a food crop (Seifu, 2014). *M. stenopetala* is a plant with a high nutritional value, drought tolerance, fast growth, and many potential medicinal uses (Abd et al., 2018). So far, the only study of *M. stenopetala* genetic diversity was conducted by Beyene (2005) in 19 accessions from Ethiopia using RAPD markers. The mean values of Nei's gene diversity and Shannon's index were 0.1818 and 0.3124, respectively, and higher values for these indexes were observed in accessions from the Konso district. This supports the hypothesis that the domestication of *M. stenopetala* occurred in Konso and then expanded to Ethiopia. Unweighted pair group method with arithmetic mean analysis (UPGMA) showed that the 19 accessions were grouped into five clusters and eight subclusters, with eight accessions grouped in the first subcluster; therefore, more polymorphic markers are required to discriminate among *M. stenopetala* accessions.

2.3.4 MORINGA OVALIFOLIA

Commonly known as phantom tree, *M. ovalifolia* is endemic from central Namibia to southern Angola. Animals consume different parts of the tree, and native people use the leaves to treat symtpoms of malaria due to their antiplasmodial activity (du Preez et al., 2017). It has been reported that *M. ovalifolia* is a sister species to *M. stenopetala, M. drouhardii*, and *M. hildebrandtii* (Olson, 2002). To test this hypothesis, Hausiku et al. (2020) investigated the intraspecific and interspecific relationships among different *M. ovalifolia* populations in Namibia using ITS sequences. The first group comprised all the *M. ovalifolia* populations from Namibia and an *M. ovalifolia* sequence retrieved from the National Center for Biotechnology Information (NCBI).

In contrast, the second group comprised sequences of *M. drouhardii, M. peregrina, M. oleifera, M. concanensis, M. hildebrandtii, M. stenopetala, M. longituba, M. borziana, M. ruspoliana, M. arborea*, and *M. rivae*. The different *M. ovalifolia* populations from Namibia were phylogenetically similar despite their morphological differences and geographic isolations, suggesting a relatively recent separation from each other. The ITS sequence data produced phylogenetic trees for *Moringa* species, consistent with Olson's phylogenetic hypothesis (2002).

2.4 ADVANTAGES AND LIMITATIONS OF MOLECULAR MARKERS USED TO ANALYZE THE GENETIC DIVERSITY IN *MORINGA* SPECIES

RAPD has been a valuable tool to assess genetic variation, particularly in medicinal plants (Yang et al., 2020). Although each technique has its advantages and limitations, the low price, speed, and simple use are certainly advantages of RAPDs versus other makers. Therefore, RAPD has been more used to investigate the genetic diversity between and within cultivated and non-cultivated genotypes of *Moringa* from different origins (Table 2.1).

However, RAPD markers presented some limitations of reproducibility and an inability to distinguish heterozygous from homozygous individuals (Garrido-Cardenas et al., 2017). For this reason, other types of molecular markers have been used to assess the genetic diversity of *Moringa*. For instance, ISSRs are the second most used markers in *Moringa* species (Table 2.1) and have been reported to be more effective than RAPDs to characterize the genetic diversity of *Moringa* as well as to distinguish among closely related genotypes (Al Khateeb et al., 2013; Saini et al., 2013; Rajalakshmi et al., 2017; Hassanein and Al-Soqeer, 2018; Hassan et al., 2020).

On the other hand, SSRs have served as useful markers to estimate the genetic diversity and population structure in *M. oleifera* (Table 2.1) because they are highly informative, codominant, multi-allele markers, in addition to being experimentally reproducible and transferable among related species (Vieira et al., 2016). Hence, SSR markers can better estimate genetic diversity compared to dominant markers (RAPD, ISSR, and AFLP). In general, the studies conducted in *M. oleifera* using SSR markers have shown high values of genetic diversity and gene flow and low genetic differentiation among the populations evaluated (Wu et al., 2010; Shahzad et al., 2013; Ganesan et al., 2014; Amao et al., 2017). Despite SSRs being extensively employed in the genetic analysis of cultivated plants, their use in orphan crops has been limited by the conventional process of construction and screening of genomic libraries (Vieira et al., 2016). To overcome this limitation, the advances made in NGS have provided a new scenario for detecting SSRs. In this regard, the information generated by the

TABLE 2.1
Studies of genetic diversity in *Moringa* species using different types of molecular markers

Moringa species	Molecular marker	No. markers tested	No. accessions tested	Collection site	References
Moringa oleifera	AFLP	4	140	India, Malawi, and Kenya	Muluvi et al., 1999
	RAPD	12	7	Tanzania	Mgendi et al., 2010
	RAPD	17	16	Brazil	Cruz da Silva et al., 2012
	RAPD	20	20	Malaysian, USA, Thailand, India, Tanzania, and Taiwan	Rufai et al., 2013
	RAPD	5	13	Nigeria	Popoola et al., 2014
	RAPD	25	8	India	Saini et al., 2013
	RAPD	10	31	Nigeria	Amao et al., 2017
	RAPD	31	7	India	Kumar et al., 2017
	RAPD	20	57	India	Panwar and Mathur, 2020
	SSR	20	24	India and Myanmar	Wu et al., 2010
	SSR	19	161	Asia, Africa, North and South America, and the Caribbean	Shahzad et al., 2013
	SSR	19	300	India	Ganesan et al., 2014
	SSR	10	31	Nigeria	Amao et al., 2017
	CyP50	11	8	India	Saini et al., 2013
	CyP50	10	7	India	Kumar et al., 2017
	CyP50	7	23	India	Ravi et al., 2020
	ISSR	10	8	India	Saini et al., 2013
	ISSR	39	97	India	Rajalakshmi et al., 2019
	ISSR	10	10	Egypt	Hassan et al., 2020
	RAM	6	45	Colombia	Chaves-Bedoya et al. 2017
	SRAP	24	97	India	Rajalakshmi et al., 2019
	SCoT	10	10	Egypt	Hassan et al., 2020
Moringa peregrine	ISSR	10	1	Jordan	Al-Khateeb et al., 2013
	ITS	2	14	Saudi Arabia	Alaklabi, 2015
	ISSR	10	7	Saudi Arabia	Hassanein and Al-Soqeer, 2018
Moringa stenopetala	RAPD	6	19	Ethiopia	Beyene, 2005
Moringa ovalifolia	ITS	2	21	Namibia	Hausiku et al., 2020

genome sequence of *M. oleifera* could be the easiest way of identifying new SSR loci (Tian et al., 2015).

2.5 CONSERVATION AND BREEDING OF *MORINGA* SPECIES USING MOLECULAR MARKERS

Plant genetic resources are the basic material for breeding new varieties that help us meet global food security. The *Moringa* genus includes economically important multipurpose trees with immense nutritional, medicinal, culinary, and phytochemical values (Abd et al., 2018). In this sense, information on the genetic diversity of *Moringa* species can provide useful data to determine conservation strategies and help breeders identify and select parental lines to establish breeding programs.

Molecular markers have provided powerful information about the genetic variation and differentiation of natural populations of *M. oleifera*. Generally, *M. oleifera* genotypes from different origins did not exhibit a specific genetic structure, which was attributed predominantly to gene flow through cross-pollination, preventing the structuration of populations and promoting the genetic diversity of the species (Muluvi et al., 1999; Mgendi et al., 2010; Wu et al., 2010; Cruz da Silva et al., 2012; Saini et al., 2013; Shahzad et al., 2013; Ganesan et al., 2014; Popoola et al., 2014; Amao et al., 2017; Kumar et al., 2017; Rajalakshmi et al., 2019).

The conservation of plant genetic resources involves several stages; first, a core collection using molecular data must be developed to maximize the representativeness of the available genetic diversity in the species and reduce plant maintenance and evaluation (Brown, 1989). However, it is strongly recommended that the information derived from molecular markers will be combined with other data, such as agronomical, morpho-phenological, nutritional, and nutraceutical traits, to maximize conservation efforts (Leone et al., 2015). Furthermore, for *in situ* conservation strategies of genetic resources, molecular markers could recognize the most representative populations within the gene pool and help establish new protected areas for their management and use (Barcaccia, 2009). Finally, the knowledge of genetic variation within local populations with fitness to different anthropological environments represents an irreplaceable gene pool that will facilitate the development of elite varieties adapted to local conditions, one of the major factors that currently limit the productivity of *Moringa* species (Leone et al., 2015).

2.6 PERSPECTIVES

Recent advances in NGS technologies have allowed the development of the genotyping by sequencing (GBS) method for SNPs discovery and assay, providing new opportunities for plant breeders with cost-effective, genome-wide scanning, and multiplexed sequencing platforms (Kim et al., 2016). GBS was originally developed by Elshire et al. (2011) to investigate the high-resolution association in maize. It is now used for many other species having a complex genome (Nadeem et al., 2018). Tian et al. (2015) reported the first high-quality genome sequence of *M. oleifera*, containing 19,465 protein-coding genes, and opened the possibility to implement the

GBS technology in the *Moringa* genus to improve our knowledge about its genetic diversity and population structure for germplasm conservation purposes. GBS could also be used in genomic selection (GS), in which their genotypes predict breeding values of individuals over multiple generations in the absence of phenotyping (Lin et al., 2014). Also, the high-density SNP panels obtained by GBS could be used in genome-wide association studies (GWAS) to identify SNP markers significantly associated with traits of interest in *Moringa*, such as the content of protein, oil, minerals and phytochemicals, tolerance to heat and drought, and resistance to plagues and diseases, among others.

2.7 CONCLUSIONS

In the last decade, numerous studies with different molecular markers have shown that there is a wide genetic diversity in wild and cultivated populations of *M. oleifera*, and despite some efforts having been made to analyze the genetic diversity and stability of *M. peregrina*, *M. stenopetala*, and *M. ovalifolia* populations, further studies are required. Most of these investigations have found more genetic variation within populations than among populations with different geographic origins, which may be due to dispersal of cuttings and seeds and high rates of gene flow between nearby populations through cross-pollination. These results suggested that provenance source is an important factor in conserving and exploiting *M. oleifera* genetic resources. However, incorporating new accessions to germplasm banks of *M. oleifera* would benefit the plant breeding programs. Finally, the recent development of genomic tools such as GBS, GS, and GWAS could be very useful to assess and exploit the wide genetic diversity available within the different *Moringa* species and to accelerate the breeding of new varieties with better agronomic performance, higher tolerance to biotic/abiotic stress, and enhanced nutritional/nutraceutical composition.

REFERENCES

Abd, N.Z., K. Husain, and E. Kumolosasi. 2018. Moringa genus: a review of phytochemistry and pharmacology. *Frontiers in Pharmacology* 9(108):1–26. https://doi.org/10.3389/fphar.2018.00108.

Adhikari, S., S. Saha, A. Biswas, T.S. Rana, T.K. Bandyopadhyay, and P. Ghosh. 2017. Application of molecular markers in plant genome analysis: a review. *Nucleus (India)* 60(3):283–297. https://doi.org/10.1007/s13237-017-0214-7.

Alaklabi, A. 2015. Genetic diversity of *Moringa peregrina* species in Saudi Arabia with ITS sequences. *Saudi Journal of Biological Sciences* 22(2):186–190. https://doi.org/10.1016/j.sjbs.2014.09.015.

Al Khateeb, W., E. Bahar, J. Lahham, D. Schroeder, and E. Hussein. 2013. Regeneration and assessment of genetic fidelity of the endangered tree *Moringa peregrina* (Forsk.) Fiori using inter simple sequence repeat (ISSR). *Physiology and Molecular Biology of Plants* 19(1):157–164. https://doi.org/10.1007/s12298-012-0149-z.

Amao, A.O., C.A. Echeckwu, D.A. Aba, M.D. Katung, and A.O. Odeseye. 2017. Diversity study of drumstick (*Moringa oleifera* Lam.) using microsatellite markers. *International Journal of Environment, Agriculture and Biotechnology* 2(5):2380–2386. https://doi.org/10.22161/ijeab/2.5.14.

Barcaccia, G. 2009. Molecular markers for characterization and conservation of crop plant germplasm. In S.M. Jain and D.S. Brar (eds), *Molecular Techniques in Crop Improvement: 2nd Edition.* Dordrecht: Springer. https://doi.org/10.1007/978-90-481-2967-6.

Beyene, D. 2005. Genetic variation in *Moringa stenopetala* germplasm of Ethiopia by using RAPD as genetic marker. MSc dissertation, Addis Ababa University. http://thesisbank. jhia.ac.ke/5873/.

Brown, A.H.D. 1989. Core collections: a practical approach to genetic resources management. *Genome* 31(2):818–824. https://doi.org/10.1139/g89-144.

Chaves-Bedoya, G., Galvis-Pérez, Z.L. and Ortiz-Rojas, L.Y. 2017. Genetic diversity of Moringa oleifera Lam. in the northeast of Colombia using RAMs markers. Revista Colombiana de Ciencias Hortícolas 11(2):408–415. https://doi.org/10.17584/ rcch.2017v11i2.7343.

Collard, B.C.Y. and D.J. Mackill. 2009. Start codon targeted (SCoT) polymorphism: a simple, novel DNA marker technique for generating gene-targeted markers in plants. *Plant Molecular Biology Reporter* 27(1):86–93. https://doi.org/10.1007/s11105-008-0060-5.

Collard, B.C.Y., MZZ Jahufer, J.B. Brouwer, and E.C.K. Pang. 2005. An introduction to markers, quantitative trait loci (QTL) mapping and marker-assisted selection for crop improvement: the basic concepts. *Euphytica* 142(1–2):169–196. https://doi.org/10.1007/ s10681-005-1681-5.

Cruz da Silva, A.V., A.R. Ferreira dos Santos, A. da Silva Lédo, R.B. Feitosa, C.S. Almeida, G.M. da Silva, and M.S. Alves Rangel. 2012. *Moringa* genetic diversity from germplasm bank using RAPD markers. *Tropical and Subtropical Agroecosystems* 15:31–39.

Dadamouny, M.A., M. Unterseher, P. König, and M. Schnittler. 2016. Population performance of *Moringa peregrina* (Forssk.) Fiori (Moringaceae) at Sinai Peninsula, Egypt in the last decades: consequences for its conservation. *Journal for Nature Conservation* 34:65–74. https://doi.org/10.1016/j.jnc.2016.08.005.

Elshire, R.J., J.C. Glaubitz, Qi Sun, J.A. Poland, K. Kawamoto, E.S. Buckler, et al. 2011. A robust, simple genotyping-by-sequencing (GBS) approach for high diversity species. *PLoS ONE* 6(5):1–10. https://doi.org/10.1371/journal.pone.0019379.

Etminan, A., A. Pour-Aboughadareh, R. Mohammadi, A. Ahmadi-Rad, A. Noori, Z. Mahdavian, et al. 2016. Applicability of start codon targeted (SCoT) and inter-simple sequence repeat (ISSR) markers for genetic diversity analysis in durum wheat genotypes. *Biotechnology and Biotechnological Equipment* 30(6):1075–1081. https:// doi.org/10.1080/13102818.2016.1228478.

Gandji, K., F.J. Chadare, R. Idohou, V.K. Salako, and A.E. Assogbadjo. 2018. Status and utilisation of *Moringa oleifera* Lam: a review. *African Crop Science Journal* 26(1):137–156.

Ganesan, S.K., R. Singh, D.R. Choudhury, J. Bharadwaj, V. Gupta, and A. Singode. 2014. Genetic diversity and population structure study of drumstick (*Moringa oleifera* Lam.) using morphological and SSR markers. *Industrial Crops and Products* 60:316–25. https://doi.org/10.1016/j.indcrop.2014.06.033.

Garrido-Cardenas, J., C. Mesa-Valle, and F. Manzano-Agugliaro. 2017. Trends in plant research using molecular markers. *Planta* 247(3):543–557. https://doi.org/10.1007/ s00425-017-2829-y.

Grover, A. and P.C. Sharma. 2014. Development and use of molecular markers: past and present. *Critical Reviews in Biotechnology* 36 (2):290–302. https://doi.org/10.3109/ 07388551.2014.959891.

Hassan, F.A.S., I.A. Ismail, R. Mazrou, and M. Hassan. 2020. Applicability of inter-simple sequence repeat (ISSR), start codon targeted (SCoT) markers and ITS2 gene sequencing for genetic diversity assessment in *Moringa oleifera* Lam. *Journal of Applied Research on Medicinal and Aromatic Plants* 18 (October 2019):100256. https://doi.org/10.1016/j.jarmap.2020.100256.

Hassanein, A.M.A. and A.A. Al-Soqeer. 2018. Morphological and genetic diversity of *Moringa oleifera* and *Moringa peregrina* genotypes. *Horticulture Environment and Biotechnology* 59(2):251–261. https://doi.org/10.1007/s13580-018-0024-0.

Hausiku, M.K., E.G. Kwembeya, P.M. Chimwamurombe, and A. Mbangu. 2020. Assessment of species boundaries of the *Moringa ovalifolia* in Namibia using nuclear ITS DNA sequence data. *South African Journal of Botany* 131:335–341. https://doi.org/10.1016/j.sajb.2020.03.002.

Kim, C., H. Guo, W. Kong, R. Chandnani, L.S. Shuang, and A.H. Paterson. 2016. Application of genotyping by sequencing technology to a variety of crop breeding programs. *Plant Science* 242:14–22. https://doi.org/10.1016/j.plantsci.2015.04.016.

Kumar, P., R. Dolkar, G. Manjunatha, and H.M. Pallavi. 2017. Molecular fingerprinting and assessment of genetic variations among advanced breeding lines of *Moringa oleifera* L. by using seed protein, RAPD and cytochrome P450 based markers. *South African Journal of Botany* 111:60–67. https://doi.org/10.1016/j.sajb.2017.03.024.

Leone, A., A. Spada, A. Battezzati, A. Schiraldi, J. Aristil, and S. Bertoli. 2015. Cultivation, genetic, ethnopharmacology, phytochemistry and pharmacology of *Moringa oleifera* leaves: an overview. *International Journal of Molecular Sciences* 16(6):12791–12835. https://doi.org/10.3390/ijms160612791.

Lin, Z., B.J. Hayes, and H.D. Daetwyler. 2014. Genomic selection in crops, trees and forages: a review. *Crop and Pasture Science* 65(11):1177–1191. https://doi.org/10.1071/CP13363.

Mgendi, M.G., M.K. Manoko, and A.M. Nyomora. 2010. Genetic diversity between cultivated and non-*cultivated Moringa oleifera* Lam. provenances assessed by RAPD markers. *Journal of Cell and Molecular Biology* 8(2):95–102.

Mosa, K.A., S. Gairola, R. Jamdade, A. El-Keblawy, K.I. Al Shaer, E.K. Al Harthi, et al. 2019. The promise of molecular and genomic techniques for biodiversity research and DNA barcoding of the Arabian Peninsula flora. *Frontiers in Plant Science* 9. https://doi.org/10.3389/fpls.2018.01929.

Muluvi, G.M., J.I. Sprent, N. Soranzo, J. Provan, D. Odee, G. Folkard, et al. 1999. Amplified fragment length polymorphism (AFLP) analysis of genetic variation in *Moringa oleifera* Lam. *Molecular Ecology* 8(3):463–470. https://doi.org/10.1046/j.1365-294X.1999.00589.x.

Nadeem, M.A., M.A. Nawaz, M.Q. Shahid, Y. Doğan, G. Comertpay, M. Yıldız, et al. 2018. DNA molecular markers in plant breeding: current status and recent advancements in genomic selection and genome editing. *Biotechnology and Biotechnological Equipment* 32(2):261–265. https://doi.org/10.1080/13102818.2017.1400401.

Olson, M.E. 2002. Combining data from DNA sequences and morphology for a phylogeny of Moringaceae (Brassicales). *Systematic Botany* 27(1):55–73. https://doi.org/10.1043/0363-6445-27.1.55.

Panwar, A. and J. Mathur. 2020. Genetic and biochemical variability among *Moringa oleifera* Lam. accessions collected from different agro-ecological zones. *Genome* 63(3):169–177. https://doi.org/10.1139/gen-2019-0102.

Poczai, P., I. Varga, N.E. Bell and J. Hyvone. 2012. Genomics meets biodiversity: advances in molecular marker development and their applications in plant genetic diversity assessment. In M. Caliskan (ed.), The Molecular Basis of Plant Genetic Diversity. London: IntechOpen. https://doi.org/10.5772/33614.

Popoola, J.O., B.O. Oluyisola, and O.O. Obembe. 2014. Genetic diversity in *Moringa oleifera* from Nigeria using fruit morpho-metric characters & random amplified polymorphic DNA (RAPD) markers. *Covenant Journal of Physical and Life Sciences* 1(2):43–60.

du Preez, I., R.-J. Bussel, and D. Mumbengegw. 2017. Evaluation of the antiplasmodial properties of Namibian medicinal plant species, *Moringa ovalifolia. Research Journal of Medicinal Plants* 11(4):167–173. https://doi.org/10.3923/rjmp.2017.167.173.

Rajalakshmi, R., S. Rajalakshmi, and A. Parida. 2017. Evaluation of the genetic diversity and population structure in drumstick (*Moringa oleifera* L.) using SSR markers. *Current Science* 112(6):1250–1256. https://doi.org/10.18520/cs/v112/i06/1250-1256.

Rajalakshmi, R., S. Rajalakshmi, and A. Parida. 2019. Genetic diversity, population structure and correlation study in *Moringa oleifera* Lam. using ISSR and SRAP markers. *Proceedings of the National Academy of Sciences India Section B – Biological Sciences* 89(4):1361–1371. https://doi.org/10.1007/s40011-018-1059-9.

Ravi, R.S. Drisya, E.A. Siril, and B.R. Nair. 2020. The efficiency of cytochrome P450 gene-based markers in accessing genetic variability of drumstick (*Moringa oleifera* Lam.) accessions. *Molecular Biology Reports* 47(4):2929–2399. https://doi.org/10.1007/s11033-020-05391-w.

Robarts, D.W.H. and A.D. Wolfe. 2014. Sequence-related amplified polymorphism (SRAP) markers: a potential resource for studies in plant molecular biology. *Applications in Plant Sciences* 2(7):1400017. https://doi.org/10.3732/apps.1400017.

Robiansyah, I., A.S. Hajar, M.A. Al-kordy, and A. Ramadan. 2014. Current status of economically important plant *Moringa peregrina* (Forrsk.) Fiori in Saudi Arabia: a review. *International Journal of Theoretical and Applied Science* 6(1):79–86.

Rufai, S., M.M. Hanafi, M.Y. Rafii, S. Ahmad, I.W. Arolu, and J. Ferdous. 2013. Genetic dissection of new genotypes of drumstick tree (*Moringa oleifera* Lam.) using random amplified polymorphic DNA marker. *BioMed Research International* 2013:604598. https://doi.org/10.1155/2013/604598.

Saini, R.K., K.R. Saad, G.A. Ravishankar, P. Giridhar, and N.P. Shetty. 2013. Genetic diversity of commercially grown *Moringa oleifera* Lam. cultivars from India by RAPD, ISSR and cytochrome P450-based markers. *Plant Systematics and Evolution* 299(7):1205–1213. https://doi.org/10.1007/s00606-013-0789-7.

Seifu, E. 2014. Actual and potential applications of *Moringa stenopetala*, underutilized indigenous vegetable of Southern Ethiopia: a review. *International Journal of Agricultural and Food Research* 3(4):8–19. https://doi.org/10.24102/ijafr.v3i4.381.

Senthilkumar, A., N. Karuvantevida, L. Rastrelli, S.S. Kurup, and A.J. Cheruth. 2018. Traditional uses, pharmacological efficacy, and phytochemistry of *Moringa peregrina* (Forssk.) Fiori. – a review. *Frontiers in Pharmacology* 9(MAY):1–17. https://doi.org/10.3389/fphar.2018.00465.

Shahzad, U., M.A. Khan, M.J. Jaskani, I.A. Khan, and S.S. Korban. 2013. Genetic diversity and population structure of *Moringa oleifera. Conservation Genetics* 14(6):1161–1172. https://doi.org/10.1007/s10592-013-0503-x.

Tian, Y., Y. Zeng, J. Zhang, C.G. Yang, L. Yan, X.J Wang, et al. 2015. High-quality reference genome of drumstick tree (*Moringa oleifera* Lam.), a potential perennial crop. *Science China Life Sciences* 58(7):627–638. https://doi.org/10.1007/s11427-015-4872-x.

Tripathy, B. and K Mallikarjunarao. 2020. Variability in tomato (*Solanum lycopersicum* L.): a review. *Journal of Pharmacognosy and Phytochemistry* 9(4):383–388.

Vieira, M.L.C., L. Santini, A.L. Diniz, and C. de Freitas Munhoz. 2016. Microsatellite markers: what they mean and why they are so useful. *Genetics and Molecular Biology* 39(3):312–328. https://doi.org/10.1590/1678-4685-GMB-2016-0027.

Wu, J.C., J. Yang, Z.J. Gu, and Y.P. Zhang. 2010. Isolation and characterization of twenty polymorphic microsatellite loci for *Moringa oleifera* (Moringaceae). *HortScience* 45(4):690–692. https://doi.org/10.21273/hortsci.45.4.690.

Yamanaka, S., E. Suzuki, M. Tanaka, Y. Takeda, J.A. Watanabe, and K.N. Watanabe. 2003. Assessment of cytochrome P450 sequences offers a useful tool for determining genetic diversity in higher plant species. *Theoretical and Applied Genetics* 108(1):1–9. https://doi.org/10.1007/s00122-003-1403-0.

Yang, M., H. Abdalrahman, S. Uwimbabazi, A.I.Mohammed, U. Vestine, M. Wang, et al. 2020. The application of DNA molecular markers in the study of *Codonopsis* species genetic variation, a review. *Cellular and Molecular Biology* 6(2):23–30. https://doi.org/10.1016/b978-0-12-812520-5.00003-1.

Yao, H., J. Song, C. Liu, K. Luo, J. Han, Y. Li, et al. 2010. Use of ITS2 region as the universal DNA barcode for plants and animals. *PLoS ONE* 5(10). https://doi.org/10.1371/journal.pone.0013102.

3 Agronomical Aspects of *Moringa oleifera* (Moringa)

Elvia Paulina Pérez-Rivera
Facultad de Ciencias Químico Biológicas, Universidad Autónoma de Sinaloa. Blvd. de las Américas y Josefa Ortiz de Domínguez, S/N, Ciudad Universitaria, C.P. 80010, Culiacán, Sinaloa, México.

Julio Montes-Ávila
Facultad de Ciencias Químico Biológicas, Universidad Autónoma de Sinaloa. Blvd. de las Américas y Josefa Ortiz de Domínguez, S/N, Ciudad Universitaria, C.P. 80010, Culiacán, Sinaloa, México.

Carlos Bell Castro-Tamayo
Facultad de Medicina Veterinaria y Zootecnia, Universidad Autónoma de Sinaloa. Blvd. San Ángel S/N, Fraccionamiento San Benito, Predio Las Coloradas, C.P. 80246 Culiacán, Sinaloa, México.

Jesús José Portillo-Loera
Facultad de Medicina Veterinaria y Zootecnia, Universidad Autónoma de Sinaloa. Blvd. San Ángel S/N, Fraccionamiento San Benito, Predio Las Coloradas, C.P. 80246 Culiacán, Sinaloa, México.

Ramón Ignacio Castillo-López
Facultad de Ciencias Químico Biológicas, Universidad Autónoma de Sinaloa. Blvd. de las Américas y Josefa Ortiz de Domínguez, S/N, Ciudad Universitaria, C.P. 80010, Culiacán, Sinaloa, México.
Corresponding author: ricastil@uas.edu.mx

CONTENTS

3.1   INTRODUCTION

Moringa oleifera is the best-known species of the genus *Moringa*, a small group of plants within the immense order Brassicales that includes the cabbage and radish families, along with the cress and caper families (APG, 2009). The family most closely related to Moringaceae is Caricaceae, with which it shares the characteristic of presenting glands at the apex of the petiole (Olson, 2002). Moringaceae comprises only the genus *Moringa*, which is composed of 13 species: *arborea, concanensis, drocanensis, drouhardii, hildebrandtii, pygmeae, peregrina, rospoliana, ovalaifolia, stenopetala, rivae, oleifera,* and *borziana* (Adams, 1972). These cover a diverse range of Habitats or growth forms, from grasses and shrubs to large trees (Olson and Razafimandimbison, 2000; Olson, 2001a, 2001b). While they vary greatly in shape, it is quite easy to distinguish a *Moringa* member from any other plant. It is a tree native to the southern Himalayas, northeast India, Bangladesh, Afghanistan, and Pakistan (Morton, 1991; Makkar and Becker, 1997). It is found throughout a large part of the planet and in Central America it is known by several common names: *Moringa,* benzolivo, mlonge, mulangay, palillo, kelor, marango, resedá, nébéday, saijhan, and sajna, among others.

The tree grows from 7 to 12 m in height, has a tuberculous root, the stem is made of soft and spongy wood, it is short and thin, with wide sloping and fragile branches (Ramachandran et al., 1980; Von Maydell, 1986). *Moringa* can be propagated through cuttings or by seeds (Palada, 1996). It is resistant to drought and tolerates annual rainfall ranging from 500 to 1500 mm. It grows in a wide range of soil types at pH 4.5–8.0 except heavy clays and prefers neutral or slightly acidic soil. The tree grows well between 0 and 1800 m.a.s.l. (Duke, 1978; Fred, 1992).

3.2 ENVIRONMENTAL ADAPTATION

3.2.1 VARIANTS

In nature, two variants of *M. oleifera* with short and long fruit pods (or just called fruits) have been reported in Poamoho, O'ahu, Hawaii (Radovich, 2011). There is great variability in the fruit size depending on the plant growing region, as well as its age and agronomic management (Resmi et al., 2005). Resmi and colleagues studied the variability between drumstick accessions of *M. oleifera* Lam. from central and southern Kerala, India, and found that accessions from different locations showed considerable variability in growth, flowering, yield, and fruit attributes. The fruit's length was observed from 32 cm (short pods) to 100 cm (long pods), as shown in Figure 3.1. The fruit circumference varied from 4.22 cm to 8.36 cm, with an average of 5.68 cm. As expected, bigger fruits had more seeds; however, these fruits were less abundant than those of moderate size.

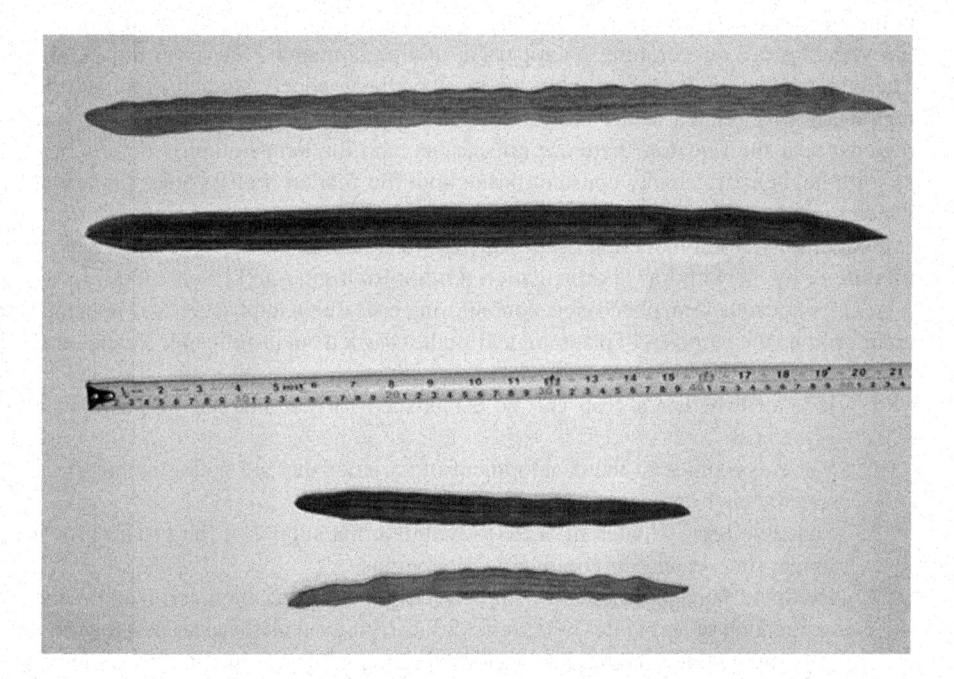

FIGURE 3.1   Mature green and dry pods from short- and long-fruited varieties of *Moringa*.

In Sinaloa, *Moringa* exists in the wild and low-tech cultivated crops with possible agronomic potential (Pérez et al., 2010). There are two variants: one corresponds to plants with fruits whose length ranges between 15 and 25 cm, known as short pod *Moringa* (SPM), while the second variant is composed of fruits of 30 and 80 cm length, called long pod *Moringa* (LPM; Castillo-Lopez et al., 2017). However, there are few studies on the nutritional and nutraceutical characterization of SPM; the vast majority, when describing the fruit of *M. oleifera*, refer to LPM, or it is taken for granted that it is this variant. Falasca and Bernabé (2008, 2009) studied the fruits of 20–45 cm length for the hybrid variety of *Moringa* PKM-1 and the extra-long fruits that can reach around 125 cm length of the hybrid variety PKM-2, but not less than 20 cm. Other authors have described wild *Moringa* fruits usually from 20 to 45 cm in length (Parrota, 1993). Castillo-Lopez et al. (2017) carried out nutritional characterization and determined the phenolic compounds of the leaves of *M. oleifera* with short and long fruit pods grown in Sinaloa, Mexico.

3.3 FLOWERING AND FRUITING

Phenology is the study of the periodic and repetitive phases or activities of the life cycle of plants and their temporal variation throughout the year. It includes the study of the causes of their synchronization and their relationship with biotic and abiotic factors, and the interrelationships that may exist between the phases of one or more species (Badeck et al., 2004). Phenological events such as sprouting, flowering, fruiting, and even senescence respond directly to macro- and microclimatic changes concerning temperature, photoperiod, solar radiation, relative humidity, and precipitation effects, being totally or partially responsible, along with genetics, for changes in the phenological states of plants (Dahlgren et al., 2007). Among the most influential variables are temperature, photoperiod, and precipitation (Sherry et al., 2007). The effects of temperature and photoperiod have been widely studied in temperate regions, mostly related to the drastic seasonal changes during the year. In tropical regions, near the Equator, there are no seasons, and the temperature is determined by altitude, being relatively constant throughout the year, as well as solar radiation, which undergoes small variations depending on the changes in cloud cover. In tropical conditions, floral induction occurs due to water stress, caused by the absence of rainfall, or by the withdrawal of irrigation (Orduz-Rodríguez and Fischer, 2007).

As mentioned before, phenological monitoring is useful in improving crop management. Among the purposes of phenological studies carried out in cultivable species are:

1. Indicate whether a crop can be established for commercial purposes in a given area.
2. Serve as a guide in the development of varieties that are better adapted to a specific environment.
3. Schedule harvest dates in order to maintain the supply of the product for a longer time, easing its commercial operations.
4. Facilitate management plans within the crop such as irrigation, fertilization, the application of herbicides or insecticides, carrying out all these tasks at the most appropriate times, taking into account the state of development of the crop.

5. Help in the interpretation or prediction of the effect that variations in the environment may have on the crop, because as suggested by Diniz et al. (2007), the organoleptic conditions in fruits can vary from one cultivar to another, depending on the genetics of the cultivar and environmental conditions such as climate and soil (Shaykewich, 1995).

Precocity may be important for plants to escape hot weather periods, the dangers of diseases and insects, or drought. The vegetative period's length can be affected by temperature, day length, altitude, soil moisture, soil fertility, and variety (Poehlman, 1976). The amplitude of the maturation period best for a production zone may not be satisfactory in another area. To complete a plant's cycle, the following phases must be considered: seed germination, the seedling phase, active growth, reproduction, and maturation (Wilson and Chester, 1981).

3.4 PHENOLOGICAL CHARACTERIZATION OF *MORINGA OLEIFERA*

Moringa oleifera is a deciduous to evergreen shrub or small tree, rarely reaching 15 m and generally less than 25 cm d.b.h. (diameter at breast hight). Initially fast-growing to 4.5 m in 9 months, the tree rarely grows more than 20 years (Schabel, 2002). It has large, spirally arranged leaves, which can reach about 60 cm in length, crowded at the end of the branches, with long petioles, pinnate in rachis. They are divided into leaflets arranged on an ovate or obovate rachis; an acute base obtuse or rounded, often oblique, apex obtuse, rounded or emarginate, whole, dull green on both sides, lighter underneath, slightly gray at first, pubescent, soon glabrous, 0.9–1.8 × 0.5–1.2 cm (Ramachandran et al., 1980; Olson and Fahey, 2011). Small glands 1 mm in length are found at the joint of each spine (Olson and Fahey, 2011).

During its flowering, it produces abundant terminal flowers on most branches. The flowers are fragrant, bisexual, oblique, pedunculated, united in erect panicles, axillary, many-flowered, densely pubescent, jointed beneath the apex, 0.7–1 cm long (Ramachandran et al., 1980; Orwa et al., 2009; Roloff et al., 2009). Calyx deeply divided into five parts, tube slightly angular, cupular-cyathiform, oblique, green, densely little pubescent on both sides, sepals of unequal size, 0.7–1.4 × 0.25–0.5 cm, five petals; unequal yellowish-white with a greenish base with fine veins, the last two and the two lateral reflections, ovate or obovate, obtuse, with a canaliculate base; on the inner side hairy at the base, the rest glabrous, 1–1.7 × 0.5–0.6 cm; the first petal erect, obovate, obtuse, glabrous inside, outside with longitudinal rows of hairs, 1.4–1.6 × 0.6–0.8 cm. Five stamens, alternating with five stabilized staminodes, densely hairy at the base, the rear stamen being the longest, 0.8–0.9 cm, the others much shorter. Ovary pedunculated, densely covered with rather long appressed hairs, terete, with three longitudinal grooves, unicellular; with three placentas, with a double row of ovules; thin style, curved white, little pubescent, hollowed at the apex (Ramachandran et al., 1980).

When the first fruits begin to mature, a new phase of vegetative growth and flower production begins (Alfaro, 2008). After flowering, they present hanging fruits, linear, acuminate, obtusely trigonal, ribbed in the form of a light capsule, woody in the

FIGURE 3.2 Flower detail (front and side view).

green stage with impressions of the seeds and half globose, and dry in the final stage of maturation. At this stage, it reaches from 10 to 30 or up to 50 cm; the fruit opens into three parts or valves; the seeds are numerous, globular, about 1–3 cm in diameter with a dark brown center, with three beige wings; wings produced at the base and apex, 2–2.5 cm long, 0.4–0.7 cm wide, scarious; the outer walls of the epidermis of the testa are thick. Under the epidermis, there is an area of parenchyma where the cell walls have numerous holes, thus presenting a reticulated appearance (Ramachandran et al., 1980; Olson and Fahey, 2011). Follows a region of fibers, up to 150 μm in length, containing crystals. The rest of the testa consists of similar parenchyma in structure to the outer zone, although the cells are longer in longitudinal section and have longer intercellular spaces. The endosperm is a single layer with oil droplets and small aleurone grains. It is associated with the aleurone layer, with 2 or 3 layers of flattened cells. The cells of the cotyledons' parenchyma also contain oil, aleurone grains, and sometimes cluster crystals. The internal cells of the cotyledons are stellate (Ramachandran et al., 1980), as shown in Figure 3.2.

It is suggested to measure these stages in degree days, so in the case of *M. oleifera*, the following durations of each of the stages have been reported (Alfaro, 2008; Alfaro and Martínez, 2008):

- Initial Phase: Seedling, germination from 5 to 18 days.
- Vegetative Phase: Vegetative development from 18 days to 8 months. Flowering from 8 to 12 months.
- Fruiting: Greater than 12 months.

Although they are genotypically determined in each variety, the succession and duration of the different stages are affected to a certain degree by environmental conditions, with climatic factors such as temperature, humidity, and the duration and intensity of light being the most important (Navarro, 1983).

3.5   EDAPHOCLIMATIC REQUIREMENTS

3.5.1   SOIL

Moringa can tolerate poor soils, drought, and salt; it requires few nutrients (Fuglie, 2001; Tshabalala et al., 2020). This may be because *Moringa* is characterized by having a tuberous root that explains its observed tolerance to the mentioned conditions (Ramachandran et al., 1980). *M. oleifera* is strictly a tropical plant and grows well in the plains; also, in almost all types of soils except rigid clays, it prefers well-drained sandy or clayey soils, tolerating clay under these conditions. It will not survive under prolonged flooding and poor drainage, as *M. oleifera* roots tend to rot in flooded soils. Soil pH is one of the factors that significantly affects the cultivation of *M. oleifera*. The tree can grow in soils with a pH between 4.5 and 8.5, but it has been observed that an optimal condition for its growth is when the soil pH is slightly acidic (6.3) to neutral (7.0) (Tshabalala et al., 2020), although there are reports that *Moringa* tolerates a soil pH of 5.0–9.0 (Ramachandran et al., 1980).

3.5.2   WATER

It is predominantly in a dry and arid area crop, where it has been found to perform well and yield profitably (Ramachandran et al., 1980). In its place of origin, it grows adequately in annual precipitation that oscillates between 750 and 2200 mm. *Moringa oleifera* is highly drought-tolerant and is produced in semi-arid and arid regions of India, Pakistan, Afghanistan, Saudi Arabia, and East Africa, receiving annual rainfall as low as 300 mm. However, such sites are likely to be irrigated or characterized by a phreatic level (Mahmood et al., 2010).

It has been reported by various authors that *M. oleifera* can grow in a wide range of annual rainfall amounts ranging from 250 to 2000 mm, although there are reports in Central America that it can receive 2500–3000 mm (Palada and Chang, 2003; Reyes, 2006; Anwar et al., 2007; Alfaro and Martínez, 2008; Radovich, 2011). In general, *M. oleifera* thrives best at altitudes below 500 m.a.s.l., although its adaptability can grow at altitudes higher than 1500 m.a.s.l., resulting in its poor growth (Palada and Chang, 2003; Olson and Fahey, 2011).

3.5.3   TEMPERATURE

Moringa tolerates a wide range of environmental conditions and can be found in tropical and subtropical climates (Palada, 1996; Olson and Fahey, 2011; Radovich, 2011; Tshabalala et al., 2020). In its native habitat, annual temperature fluctuations tend to be very large, with minimum and maximum shade temperatures ranging between 1 and 3°C and between 38 and 48°C during the colder and warmer months, respectively

(Mahmood et al., 2010). It grows best between 25 and 35°C but tolerates up to 48°C in the shade and can survive a light frost. Growth slows significantly at temperatures below 20°C (Palada and Chang, 2003; Radovich, 2011).

In Mexico, it has been observed that the most suitable average temperatures for its growth are in the range of 26–29°C. Different combinations of conditions can damage plant growth, being below 10°C; in these cases, low temperatures damage only the terminal branches, but in others, only the trunk survives. In growing conditions of high temperatures above 40°C, *M. oleifera* seems to adapt despite lacking the ideal climate, and a possible explanation is that the rain contributes to its adequate acclimatization (Olson and Alvarado-Cárdenas, 2016).

3.6 PARTICULARITIES OF THE CROP

3.6.1 CROP PROPAGATION METHODS

Plant propagation techniques to produce a high-quality seedling are important in developing a perennial crop, including *Moringa* (Santoso and Parwata, 2020). The tree can be propagated from seeds or cuttings (Kiragu et al., 2015). In India, stem cuttings that root relatively easily are generally preferred. Plants that are grown from seeds produce lower-quality fruits (Ramachandran et al., 1980). This may be due to poor soils and important deficiencies, and inequalities in water access (Gandy, 2008).

The stem cuttings should be large, planted in moist soil, easily take root, and grow into sizeable trees within a few months. Large branch cuttings 1–1.35 m in length and 14–16 cm in circumference are planted *in situ* in the rainy season (June–August). In the Salem district of Tamil Nadu (India), trees are felled, leaving a stump from which 1–2 shoots are allowed to grow. From these shoots, cuttings 2 m long and 4–5 cm in diameter are selected and used as planting material (Ramachandran et al., 1980).

In the Kanyakumari district of Tamil Nadu, budding from the gusset graft has proven successful, with budding trees beginning to produce in 6 months and continuing to yield a good harvest for approximately 13 years. The season between September and December is the best for bud break. *Moringa* is planted 3–5 m apart in both directions, and irrigation is done in the initial stages until the plants are well established. Irrigation and compost are rarely practiced, but in Kerala, ring ditches are dug around the trees and green leaves, barnyard manure, and ash are placed in the ditches about 10 cm from the tree during the rainy season. This practice is said to encourage high returns (Ramachandran et al., 1980).

Direct seeding is preferred when there is a lot of seed available and labor is limited. Transplantation allows flexibility in planting in the field but requires additional work and costs to grow seedlings. Stem cuttings are used when seed availability is limited, but labor is abundant (Palada and Chang, 2003).

The seeds can be planted directly in the ground or in containers (plant pots). They do not require any treatment and germinate rapidly in the first 6–10 days after sowing and can reach up to 5 m height in a year under controlled environmental conditions if new seeds are used. However, the percentage decreases as the time of obtaining these passes. *Moringa* seeds do not have dormancy periods and can be planted as soon as they are ripe (Padilla et al., 2012).

The cultivation of *M. oleifera* through seeds has better results than cuttings under equal cultivation conditions. Propagation of *M. oleifera* by seeds or cuttings is not apparent in their growth and survival performance when they have similar irrigation conditions. This was observed in the seedlings and cuttings grown in the greenhouse that performed better than normal conditions. If cuttings are to be used in the propagation of *M. oleifera*, cuttings of the 14–16 mm diameter class will give better results. This is because larger diameter cuttings can continue to support sprouting branches before proper root development is achieved to allow cuttings to utilize soil nutrients (Kiragu et al., 2015).

3.6.2　PLANTATION

3.6.2.1　Seed Selection

The seeds are selected considering some important variables, either according to the farmers' experience in the field or some considerations reported in *Moringa* cultivation technical guides (Alfaro and Martínez, 2008; Reyes and Mendieta, 2017). For example, new seeds, that is, recently collected seeds from the central part of larger pods that are generally the largest and heaviest seeds, and the brightness of the seed (Padilla et al., 2012; Valdés-Rodríguez et al., 2018), as shown in Figure 3.3.

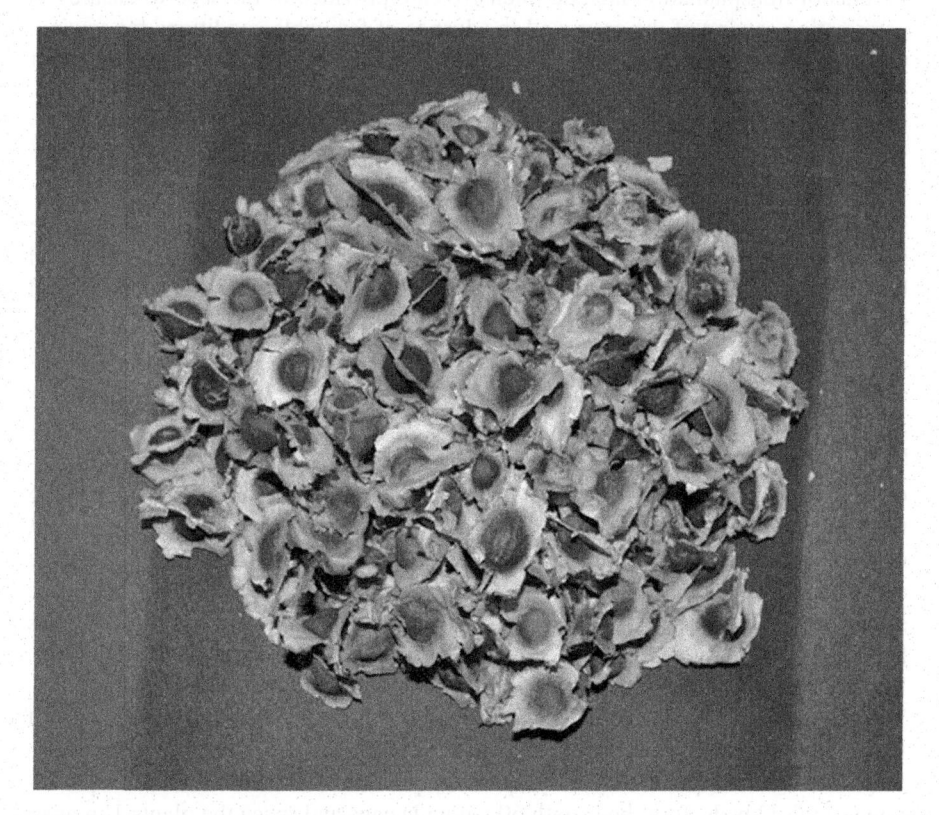

FIGURE 3.3　Dark brown seed with three thin wings.

In northern Indian conditions, the tree loses its leaves in December–January, and new leaves appear in February–March. Flowers and tender whip-shaped fruits follow it during April–June, which ripen during the summer (Ramachandran et al., 1980). It is during this time that the best trees for seed production can be identified. On the other hand, in Mexico and Central America, it is possible to locate the best seed-producing trees during August, according to the development and abundance of flowers. The production time of the seed is from October, when it begins to mature, extending until April of the following year (Alfaro and Martínez, 2008).

Germination is commonly between 60% and 90% for fresh seeds and occurs between 7 and 30 days after sowing. The seeds do not retain their viability stored at room temperature for more than 2 months (Parrota, 2004); the germination rate can decrease from 10% to 52% after one month of storage, while some selections have shown germination of 25% to 60% after 12 months. No seed of this species will germinate after 24 months (Morton, 1991).

Direct sowing is recommended as long as there are conditions for the control of insects and the availability of the seed is abundant (50% more than is needed) to compensate for the losses that may occur. Otherwise, it is better to opt for nursery preparation (Alfaro and Martínez, 2008). Uprooting the seedlings from the nursery to the prepared field with the recommended spacing is known as transplantation. The benefits of transplantation include intensive care provided to the seed to ensure the germinating sowing of the crop well in advance of favorable conditions for sowing, the availability of viable, healthy seedlings, and the maximum establishment of seedlings. Transplantation allows flexibility in planting in the field but requires additional work (Olorukooba et al., 2013).

3.6.2.2 Seed Activation

It is proven that when *Moringa* seeds are soaked under running water for 24 h before sowing to accelerate germination, they show an excellent germination index and better development compared to those that did not receive this treatment. The soaked seed offers a softened testa (Padilla et al., 2012; Montilla-Mota et al., 2017).

3.6.2.3 Planting Methods

Method 1. Direct Sowing

Trees are easily grown from seed, and direct seeding is the most common commercial production method in India (Radovich, 2011). Direct seeding involves placing the seed directly into the prepared soil. It can be done mechanically, using a seeder, or manually if the seeder is not available (Reyes and Mendieta, 2017). Two or three seeds are sown per loin at a depth of 2 cm; 2 weeks after germination, thin to the strongest seedling per hill (Palada and Chang, 2003). The density of 1 million plants ha^{-1} has been considered optimal due to fresh biomass production, planting cost, harvesting, and weed control in good agroclimatic conditions (Foidl et al., 1999). To produce leaves, pods, and seeds, place the plants at 3–5 m between the rows. If raised beds are used, form beds with 2 m wide tops and space the plants 3–5 m in a single row. For leaf production only, space the plants 50 cm within rows 1 m apart. If you use raised beds, form beds with 60 cm wide caps and space the plants 1 m apart

FIGURE 3.4 Wide rows are used for leaf, pod, and seed production.

in a single row. For intensive leaf production, plant plants 10–20 cm apart in rows 30–50 cm apart. Closer spacing allows harvesting of young edible shoots every 2–3 weeks (Palada and Chang, 2003), as shown in Figure 3.4.

To produce seed exclusively, sowing at a distance of 2 m × 2 m stands out with an average yield of three years of 752.37 kg ha^{-1} seed (Basulto Graniel et al., 2017), although other authors recommend that sowing densities of 3 m × 1 m and 3 m × 2 m obtain the highest seed yields (Ledea-Rodríguez et al., 2018). Alternatively, for pod production, the recommended spacing is 2.5 m × 2.5 m (Radovich, 2011).

Method 2. Transplantation

Uprooting the seedlings from the nursery to the prepared field with the recommended spacing is known as transplanting. The benefits of transplantation are counting the availability of viable, healthy seedlings and the maximum establishment of seedlings. Transplantation allows flexibility in planting in the field. Seedlings can be grown from seed in divided trays, individual pots, plastic bags, or seedlings. The use of divided trays and separate containers is preferred because there is less damage to the seedlings when transplanted (Palada and Chang, 2003; Olorukooba et al., 2013). The seeds in these implements are sown in the commercial substrate (horticultural grade peat substrate), or they can be prepared from a substrate made with 60% sand and 40% black soil in such a way that the soil texture is sandy loam. This can be achieved, for example, with a mixture made with a part of black soil, one of sand, and one of organic matter, previously sifted. The seedlings need to reach 15 cm in average height for their transplantation in soil or bags (Alfaro and Martínez, 2008; Carranco et al., 2016).

Method 3. Using Stem Cuttings

Compared to trees planted from seeds, trees from stem cuttings grow faster but develop a shallow root system that makes them more susceptible to moisture stress and wind damage. Cuttings are also more sensitive to termite attacks (Palada and Chang, 2003; Saint Sauveur and Broin, 2010). *Moringa* cuttings cut at the end of the dry season, that is, at the beginning of the rainy season, show a 95% seizure and a 90% survival rate. The cuttings are left to root with their reserves to obtain these high percentages. Later they are transplanted to the final soil, which must have a good humidity regime. Once the cuttings have been received, a good practice for rooting is to place them vertically under shade and bury them about 10 cm in the ground (Ramachandran et al., 1980; Reyes and Mendieta, 2017).

3.6.3 SOWING

The selection of the sowing moment is a key element in the establishment process. The optimal season varies according to locality, but in general terms, sowing should be done during the rainy period, when the soil moisture is adequate for germination and establishment of the crop (Reyes and Mendieta, 2017). However, in areas with irrigation, it can be sown at any time of the year, as long as the frequency and volume of water application are guaranteed to allow the plants' germination and survival during the establishment phase. Some technicians and producers in Cuba who have irrigation prefer to plant in the dry season to control the weeds, pests, and diseases that attack plants in their young stages (Padilla et al., 2017). Depending on the soil's characteristics, picking and tracking will be carried out (double or triple); this work should preferably be carried out at the end of the dry season or the beginning of the rainy season (Pérez et al., 2010).

Sowing can be done both by botanical seed and by cuttings. This will depend on various factors, including the availability of seeds or conditions for cultivation in nurseries. In certain areas, flowering is difficult due to weather conditions, and then planting must be done by stakes or by plants previously developed in nurseries. The trees obtained from seeds produce stronger and deeper roots (e.g., arid and semi-arid regions); therefore, it is convenient to plant by botanical seeds (Kiragu et al., 2015; Padilla et al., 2017), as shown in Figure 3.5.

If transplantation is chosen, it should be done in cool hours. The afternoon is recommended, thus avoiding stress. During transplantation, it should be considered that winter is fully established with frequent rains. The heatwave season, which occurs at the end of July and during August, should be considered, as well as the frequency of unstable rain; this will avoid losses due to insolation or water stress. It is advisable to water the plants one day before transplanting. In clay and rocky soils, the plant can adapt and develop to such a degree that they already produce flowers and the first pods after 9 months. In very adverse soil conditions, organic matter or composted manure can be added to the soil (Alfaro and Martínez, 2008).

FIGURE 3.5　*Moringa oleifera* cultivation in Mojolo, Culiacán, Sinaloa, México.

The number of seeds, seedlings, or cuttings to be used will depend on the density of plants wanted in the field and factors such as the weight and viability of the seed, the distance between rows, the survival of the seedlings or cuttings, and environmental conditions. An adequate plant density is key to obtaining maximum biomass production, due to the greater use of solar radiation through greater foliar coverage. Additionally, there is adequate root development at higher population densities, increasing the absorption of nutrients and retaining moisture in the soil (Alvarado-Ramírez et al., 2018).

According to Reyes et al. (2006), a density of 750,000 plants ha^{-1} increases the biomass yield in dry matter (DM); one million plants ha^{-1} is the optimal population density, considering the biomass production, costs of establishment, cutting management, and weed control (Foidl et al., 2001).

Although according to Godino García (2016), the planting density will depend on the purpose of the crop, they recommend densities higher than 30,000 plants ha^{-1} to produce biomass for animal feed. Its life as a crop ranges from 4 to 10 years. The cuts are made, the first at 60 days, and the rest at 45 days in the dry season and 30 days in rainy or irrigated conditions. For the production of seeds and oil, they recommend a cultivation density of 300–1200 plants ha^{-1}, recommending not to harvest the first year and dedicate it to the tree top formation. The average production is 4500 kg of seed per hectare, which increases until the fourth yield stabilizes, remaining constant for 15–20 years. A cultivation density of 1000–10,000 plants ha^{-1} is adequate for producing pods for human consumption. The harvest of fresh pods begins 3 months after transplantation (5 if it is sown direct). In India, pod production is 19 kg tree^{-1}

year^{-1}, although in Hawaii, selected varieties allow for production between 3 and 8 times greater. Finally, to produce leaves for human consumption, the recommended cultivation density is 1000–10,000 plants ha^{-1} with manual harvest; each tree grows from 1 to 5 kg of fresh leaf annually.

3.6.4 COMPOST AND FERTILIZATION

Moringa can be established well in most soils without fertilizers, as its broad and deep root system allows the efficient absorption of the nutrients present in the soil. However, as *Moringa* is a tree with the ability to generate high volumes of biomass, it must be considered that high productivity implies large extraction of nutrients from the soil. Therefore, a fertilization program is necessary to maintain stable production over time (Padilla et al., 2017).

Research work at the Tamil Nadu Agricultural University (India) has shown that the application of 7.5 kg of free-range manure and 0.37 kg of ammonium sulfate per tree (in December) tripled the yield compared to non-fertilized trees (Ramachandran et al., 1980). Other reports recommend digging trenches around the plant's base (10–20 cm from the base) and applying approximately 300 g of commercial nitrogen fertilizer per tree. If commercial nitrogenous fertilizers are not available, use compost or well-rotted free-range manure at a rate of 1–2 kg tree^{-1} (Palada and Chang, 2003). If chemical fertilization is chosen, the application needs will depend on the previous soil analysis. Nevertheless, it can handle a base application of 200 kg of urea and 200 kg of a soil fractionated mixture (soil, compost and minerals) in two applications. In the first application (pre-sowing), 200 kg of complete is applied, and after the first pruning, 200 kg of urea is used (Reyes and Mendieta, 2017).

The variation in the amounts of fertilizers may be due to the type of soil where the cultivation is carried out; therefore, it is recommended before the work that soil analysis can be carried out to know the state of the nutrients in the soil so that adequate fertilization is provided (Pérez et al., 2010).

3.6.5 PRUNING

Moringa tolerates regular pruning; its shoot emission is 4–5 branches per stem on average, with higher branch growth when the stem is pruned between 6 and 8 cm in diameter (Casanova-Lugo et al., 2018). This should be done to promote branching, increase yields, and facilitate harvesting. If the main stem is left to grow without cutting, the fast-growing tree will grow straight and tall and produce leaves and pods only on the main stem (Palada and Chang, 2003).

To stimulate the development of many branches and pods within easy reach from the ground, prune the shoot for apical growth when the tree is between 1.0 and 2.0 m tall (approximately in September). Only the stem is left (at 80 cm high) to provoke the formation of sprouts; 3 months later (when the shoots reach approximately 80 cm long), a second pruning is carried out (Pérez et al., 2010). New shoots will emerge just below where the cut is made. After that, cut the growing tips off the branches to make the tree leafier. Another pruning strategy is to trim each branch 30 cm when it reaches 60 cm in length. This will produce a multi-branched shrub (Palada and Chang, 2003; Olorukooba et al., 2013; Zaku et al., 2015; Ledea-Rodríguez et al., 2018).

If the tree is being grown for pod production, remove the flowers for the first year. This will channel all the energy from the sapling into vegetative and root development (rather than energy-draining pods), leading to more vigorous growth and productive yields in the future. Older trees that are unproductive or too tall for easy harvesting can be pruned to ground level. New shoots will quickly emerge from the base (Palada and Chang, 2003).

Light and moderate pruning are sufficient to promote further canopy development, resulting in better flowering than severe pruning. Plants pruned at 1 m receive better initial penetration of light into the plant architecture; plant growth was not sufficient to outperform trees pruned at 2 m above the soil surface. The production of *M. oleifera* leaves from moderately pruned trees is promising when grown under suboptimal growing conditions (Du Toit et al., 2020).

3.6.6 HARVEST

3.6.6.1 Foliage Harvest

In the vast majority of places where *Moringa* is grown, it has similar behavior to its place of origin, India, regarding the production of flowers and fruits. As they appear twice a year, there are two harvests from July to September and March to April (Ramachandran et al., 1980).

The leaves can be harvested after the plants grow to 1.5–2.0 m, which usually takes at least a year. Harvest the leaves by breaking the stems from the branches. Harvesting young shoot tips will promote lateral branches' development where cuts are made along the main branches. Allow plants to develop new shoots and branches before subsequent harvests (Palada and Chang, 2003).

The frequency with which the foliage of *M. oleifera* is cut is a factor that affects the plant's behavior, reflecting variation in biomass production, as shown by the experiment carried out by Reyes et al. (2006). They evaluated stages to harvest forage of 45, 60, and 75 days of regrowth, registering dry matter of 9.1, 11, and 17.6 t ha^{-1} year^{-1}, respectively (Carranco et al., 2016).

3.6.6.2 Pods

The first time *Moringa* bears fruit, the yield is generally low in the first 2 years, but from the third year on, a single tree produces 600 or more fruits per year (Ramachandran et al., 1980). When harvesting pods for human consumption, harvest when the pods are still young and break easily. Older pods develop a hard exterior, but the white seeds and pulp remain edible until the ripening process begins. When producing seeds for planting or oil extraction, allow the pods to dry out and turn brown on the tree. In some cases, it may be necessary to prop up a branch that contains many pods to prevent it from breaking. Harvest the pods before they open and the seeds fall to the ground. The seeds can be stored in well-ventilated bags in dry and shady places (Zaku et al., 2015).

3.6.6.3 Seed

Trees generally produce good seed crops for about 12 years (Roloff et al., 2009). Pods are normally produced during the second year of growth. Harvest the pods before they split open and fall to the ground. Store seeds in well-ventilated bags in a cool, dry, shady area. The seeds are still viable for planting 2 years later (see more information in Section 3.6.2.1 on seed selection; Palada and Chang, 2003).

3.7 LEAF QUALITY PARAMETERS

The *Moringa* leaf is becoming a first-rate resource, with low production cost, which has shown nutritional and nutraceutical richness with its main use as a food supplement. In recent years, this plant's growth has gained interest due to scientific discoveries highlighting its nutritional and medicinal properties (Fuglie, 2001; Fahey, 2005; Ferreira et al., 2008; Olson and Fahey, 2011). Its usefulness as a non-food product has also been widely described (wood, charcoal, lubricating oil, water clarifier, etc.; Foidl et al., 2001; Falasca and Bernabé, 2008; Muñoz et al., 2008). In many Asian and African countries, it has been traditionally used as food for humans and animals (Pérez et al., 2010). This tree has great potential for cultivation in Mexico and many parts of tropical America for its unique properties.

One of the most attractive features of *Moringa* is the high protein content in its leaves. The analysis of the dried leaves' protein content shows that up to 30% of their weight is made up of protein (powdered milk contains 35%) and that most of this seems to be directly assimilable. Furthermore, the leaves contain all essential amino acids (the units of proteins that the body cannot synthesize) in a high and well-balanced profile (Freiberger et al., 1998). For all this, MorinIga is an important food source, a finding that has been proven repeatedly (Richter et al., 2003). Many plants display structures rich in protein, for example, several pulses (beans). However, while most of them produce these proteins in their fruits, *Moringa* stands out for containing the proteins in its leaves, which are present on the tree practically all year round.

Even though *Moringa* is characterized by its high content of proteins and vitamins, it contains very low antinutritive substances that are found in nonlethal doses; furthermore, some substances that cause secondary or adverse effects are found in insignificant quantities (Makkar and Becker, 1997; Olson and Fahey, 2011). Makkar and Becker (1996) showed that *Moringa* leaves contained negligible amounts of tannins; likewise, their analyses showed no evidence of lectins or trypsin inhibitors. Saponins were found, but in low quantities, equivalent to the levels recorded in soybeans, at innocuous levels and no hemolytic activity was found (Makkar and Becker, 1997; Gidamis et al., 2003). Phenolic compounds are among the nutraceutical compounds present in *Moringa*, such as benzoic acids, zeatin, quercetin, beta-sitosterol, caffeoylquinic acid, kaempferol, and mainly benzyl isocyanate (Atawodi et al., 2010).

3.8 CROP PESTS AND DISEASES

Moringa oleifera is resistant to most pests and diseases, but outbreaks can occur under certain conditions. For example, diploidy root rot can appear in flooded soils, causing severe wilting and plant death (Palada and Chang, 2003). According to Ramachandra et al. (1980), *M. oleifera* is not affected by any serious disease in India, although root rot caused by *Diplodia* sp. was reported in Madras, India. The hairy caterpillar (*Eupterote mollifera* Wlk) causes defoliation in *Moringa*. Other insect pests on *Moringa* include: caterpillars of *Tetragonia siva, Metanastia hyrtaca*, and *Heliothis armigera*; an aphid (*Aphis caraccivera*); a scale insect (*Ceroplastodes cajani*); a borer (*Diaxenopsis apomecynoides*); and a fruit fly (*Gitonia* sp.). Palada and Chang (2003) report that, in plantations in Shanhua, Taiwan, mite populations can increase during dry and cool weather. These pests cause leaves to turn yellow, but plants generally recover their green color during hot weather. Other insect pests include termites, aphids, leaf miners, whiteflies, and caterpillars: *Atta mexicana, Corythucha gossypii*, and *Sagotylus confluens* (Valdés-Rodríguez et al., 2018).

In Sinaloa (Mexico), the most common pests reported for *M. oleifera* are mochomos (*Atta* sp.) and armyworm (*Heliothis zea*), and hairy (*Estigmene acrea*) and meter (*Trichoplusi ni*) worms. Some other pests and diseases attack the crop, especially in winter and spring, such as stem screwworm (*Diatraea* sp.) and whitefly (*Bemisia* sp.); the infection caused by the latter is not serious and in general the plant recovers on its own. Some fungi that cause stem and root rot (*Phytophthora* sp.) could affect *Moringa* seedlings; therefore, it is necessary to be aware of the appearance of any symptoms, such as wilting, yellowing, and leaf drop or rotting of the base of the stem. If any of these signs appear, it is necessary to consult a plant specialist because various fungi can produce similar symptoms (Perez et al., 2010). Similar infections to those previously exposed have been reported in Hawaii (Radovich, 2011). Occasionally, severe damage to tree trunks by borers has been observed. In eastern O'ahu, extinction-like symptoms have been observed in *Moringa* seedlings.

3.9 AGRONOMIC AND NUTRITIONAL ADVANTAGES OF *MORINGA* OVER OTHER CROPS

In the short and medium term, *Moringa* brings together agronomic and technological advantages such as adaptation to various environments, low water requirements, high potential for grain and foliage yield, and high oil content and quality of this versatile species. It has also gained great interest due to its nutritional and energetic qualities, which is why it is positioned as a production alternative in various regions of the world, such as Asia and Africa (Basulto Graniel et al., 2017). This may be possible because the species has great ecological plasticity because it can adapt to various soil and climate conditions (Pérez et al., 2010).

One of the characteristics that have attracted the most attention is the high content of protein in its leaves, constituting up to 30% of its dry weight (Garavito, 2008; Olson and Fahey, 2011), being ideal for feeding different animals: cattle, sheep, pigs, fish, and birds (Basulto Graniel et al., 2018). For this reason, its use as forage is of great importance due to its good nutritional characteristics and its high yield of fresh biomass production (Foidl et al., 1999; Pérez et al., 2010).

Due to the above, there are several comparative studies of *Moringa* yields against other crops. It presents high productivity of green matter compared to other grasses, such as alfalfa, and the highest values are reached with a planting density of one million plants per hectare (Makkar and Becker, 1996). These studies are derived from the fact that animal nutrition is compromised in dry times, and forage or food sources become more expensive. Some organic flours such as soybeans, cottonseed cake, and grasses are being used to overcome forage shortages during these periods. However, these have some limitations, such as unavailability from December to May, as currently green forage is less available after the harvest of wheat, alfalfa, brassica, and corn. This leads to reduced livestock production. Relatively low irrigation requirements make *Moringa* superior to some other livestock flours such as soybeans, cottonseed cake, and grasses, which require fairly high irrigation to avoid a reduction of livestock production. Soybeans, for example, require intensive irrigation, which makes it difficult for small farmers to grow them (Nouman et al., 2014).

As explained in previous sections, *Moringa* can have two pruning or harvest cultural practices in the year. However, some authors report that it can be generalized that it is possible to obtain four cuts of *M. oleifera* forage per year in the lower parts of the state of Nuevo León (Mexico). Here, an average fresh forage yield of 134.8 t ha^{-1} year^{-1} has been reported, with optimal irrigation conditions or good rainfall during the year (Carranco et al., 2016).

In addition, *Moringa* can be competitive against alfalfa in those climates whose average annual temperature is above 18.7°C and the average temperature of the months with a vegetative period is between 25 and 35°C (alfalfa requires average temperatures of 15°C and an optimal range of 18–28°C; Godino García, 2016).

Most of the grasses used present forage yields of 2–5 t DM ha^{-1}, low nutritional value (around 40% digestibility), and crude protein contents between 3$ and 7%. In contrast, *Moringa* contains 321–521 g kg^{-1} DM of neutral detergent fiber, 224–361 g kg^{-1} DM of acid detergent fiber, approximately 2.27 Mcal kg^{-1} DM metabolizable energy, and 79% *in vitro* digestibility of dry matter. Likewise, the presence of 19 amino acids has been reported, of which 10 are essential in animal nutrition; leucine and methionine are found in greater and lesser proportions with 19.6 and 2.9 mg g^{-1} of DM, respectively, similar to that reported for soybeans (Reyes-Sánchez and Mendieta Araica, 2017; Casanova-Lugo et al., 2018; Ramírez et al., 2020).

The edible fraction of *M. oleifera* is higher in crude protein content and lower in fibrous components as compared to other forages such as buffelgrass (*Cenchrus ciliaris* L.), pretoria (*Dichanthium annulatum* (Forssk.) Stapf), and Johnson grass (*Sorghum halepense* (L.) Pers.). Therefore, *Moringa* can be considered one of the most important crops globally, with a high impact on animal nutrition (Casanova-Lugo et al., 2018).

The effects of the previous composition in *Moringa* were evaluated in sheep diets, highlighting that the protein digestibility was higher than that of alfalfa (79.8% versus 75% for alfalfa), which means that *Moringa* proteins are more digestible than those of alfalfa (Pérez et al., 2010). In the investigations of the use of *M. oleifera* as fresh forage for dairy cattle, in animals grazing and supplemented with concentrate and subsequently passed to grazing and *Moringa* supplement, no decrease in milk volumes and no palatability problems have been found. In these experiments, the cost of *Moringa* is 10% concerning the concentrate (Foidl et al., 1999).

Rivero et al. (2020) studied the productive behavior of growing pigs fed with fresh foliage of *M. oleifera* Lam. in partial replacement of soybean cake and corn. The feed conversion response was positive, requiring a reduced amount of feed as soy was replaced by fresh *Moringa* foliage in the diet. Castillo-López et al. (2018) supplemented feed for Japanese quail (*Coturnix japonica*) with a *Moringa* leaf meal, showing no significant weight gain differences and feed conversion compared to the traditional soybean meal. Additionally, they found a similar effect to growth-promoting agents (GPA) without causing hematological damage in Japanese quail during the fattening period. In conclusion, *Moringa* can be used as feed *par excellence* as a protein source for the fattening of farm animals.

These results provide an alternative for producers throughout the production chain, from animal feed to its application in animal husbandry whatever its purpose, such as fattening, source of other products, maintenance, and health, among others. Therefore, it is not only about reducing food costs or compromising the quality of the final product with the alternatives used until today, but rather producers can adopt this alternative that adds value to their products and guarantees their consumers almost zero risks when consuming them. Additionally, these actions should focus directly on animals' welfare with better breeding conditions, fattening, and products derived from them.

3.10 ASPECTS AND ECONOMIC IMPORTANCE OF THE *MORINGA* CULTIVATION

The 13 species of *Moringa* are currently of great interest for their extraordinary economic potential. The few species studied provide nutrients for leafy vegetables and fruits, high-quality seed oil, antibiotics, and water clarifying agents (Olson and Razafimandimbison, 2000). The most widely distributed and cultivated worldwide is *M. oleifera* due to its multipurpose nature (Olson and Fahey, 2011; Estrada-Hernández et al., 2016). This is due to its relatively easy propagation, spreading through sexual and asexual means, and its low demand for nutrients from the soil and water after planting, facilitating its production and management. Introducing this plant to a farm that has a biodiverse environment can be beneficial to both the farm owner and the surrounding ecosystem (Fuglie, 2001).

In Mexico, as in other parts of the world where *Moringa* is grown and sold, the main reasons for establishing plantations are related to its use for family health and nutrition, followed by forage potential in livestock, and finally as a productive alternative. All *Moringa* crops are associated with other crops, such as corn, beans, forest species, and fruit trees. All *Moringa* producers consume their *Moringa*; 83% sell it for human consumption, while 20% sell it or use it for livestock feed. All producers commercialize the leaf, and only 27% commercialize the seed; 97% of producers use organic fertilizers and agroecological practices (Paliwal et al., 2011; Mota-Fernández et al., 2019).

In summary, *M. oleifera* is a source of materials that can have multiple uses. The scientific evidence of the properties of *M. oleifera* makes this plant a firm candidate in the search for medicinal, chemical, industrial, agricultural, and nutritional alternatives that satisfy the specific deficiencies of each sector; placing this species as one of the strategic crops for the development of farming activities in tropical and subtropical zones. *Moringa oleifera* is a multipurpose plant that can be of interest in most industrial sectors and made available for humanity's benefit due to its comprehensive use.

However, it is still very important to develop experimental research to verify and expand the information and knowledge are available so far, as this type of study could bring solutions to the serious problems of malnutrition, contamination, diseases, among others that we face today (Olson and Fahey, 2011; Gómez and Angulo, 2014; Mota-Fernández et al., 2019).

3.11 FUTURE AGRO-INDUSTRIAL APPLICATIONS

Obtaining raw material to produce bioenergetics is the limiting factor for its production. This lack of raw material for the national industry drives new oil-producing crops for biofuels production. *Moringa oleifera* can be a viable option for the generation of biofuel, with agronomic and technological advantages such as high potential for seed and foliage yield, as well as high oil content and nutritional quality of the leaves (Basulto Graniel et al., 2017). Therefore, it not only has the potential to be a source of oil for the biofuels production, but it could also be possible to produce functional and nutraceutical foods due to the more recent characterization of the quality of its leaves and its phytochemical components; for instance, its antioxidant and antiproliferative molecules.

3.12 CONCLUSION

Moringa oleifera is presented as a multipurpose renewable forage, nutritional and energy alternative from which raw materials can be developed due to its food value for humans, animals, and bioenergetic crops. However, standardized agronomic management and agro-industrial performance are unknown; therefore, it is necessary to start selecting elite genotypes and developing sustainable production technologies for *Moringa* as a source of animal and human food, as well as raw material for agro-industrial use.

REFERENCES

Adams, C. D. (1972). *Flowering Plants of Jamaica*. Mona: Universities of the West Indies).

Alfaro, N. C. (2008). Rendimiento y uso potencial de Paraíso Blanco, *Moringa oleifera* Lam en la Producción de alimentos de alto valor nutritivo para su utilización en comunidades de alta vulnerabilidad alimentario-nutricional de Guatemala, Consejo Nacional de Ciencia y Tecnología – CONCYT – Guatemala, Informe Final. Guatemala: FODECYT. www.sica.int/documentos/uso-potencial-de-la-moringa-moringa-oleifera-lam-para-la-produccion-de-alimentos-nutricionalmente-mejorados_1_36997.html.

Alfaro, N. C. and Martínez, W. (2008). Uso potencial de la Moringa (*Moringa oleifera* Lam) para la producción de alimentos nutricionalmente mejorados. Instituto de Nutrición de Centroamérica y Panamá-INCAP. Guatemala: FONACYT. www.sica.int/download/?36997.

Alvarado-Ramírez, E. R., Joaquín-Cancino, S., Estrada-Drouailet, B., Martínez-González, J. C., and Hernández-Meléndez, J. (2018). *Moringa oleifera* Lam.: Una alternativa forrajera en la producción pecuaria en México. *Agroproductividad*, 11(2): 108.

Anwar, F., Latif, S., Ashraf, M., and Gilani, A. H. (2007). *Moringa oleifera*: a food plant with multiple medicinal uses. *Phytotherapy Research*, 21(1): 1725.

Atawodi, S. E., Atawodi, J. C., Idakwo, G. A., Pfundstein, B., Haubner, R., Wurtele, G., et al. (2010). Evaluation of the polyphenol content and antioxidant properties of methanol extracts of the leaves, stem and root barks of *Moringa oleifera*. Lam. *Journal of Medicinal Food*, 13(3): 710–716.

APG (Angiosperm Phylogeny Group). (2009). An update of the Angiosperm Phylogeny Group classification for the orders and families of flowering plants: APG III. *Botanical Journal of the Linnean Society*, 161(2): 105–121.

Badeck, F., Bandeau, A., Böttcher, K., Doktor, D., Lucht, W., Schaber, J., et al. (2004). Responses of spring phenology to climate change. *New Phytologist*, 162(2): 295–309.

Basulto Graniel, J. A., Gamboa, J. A., and Reyes Chávez, E. (2017). Densidades de siembra de moringa en Yucatán, México. *Seguridad Alimentaria: Aportaciones Científicas y Agrotecnológicas*, 405.

Basulto Graniel, J. A., Burgos Díaz, J., and Gamboa, J. (2018). Producción de biomasa de 20 colectas de moringa bajo las condiciones agroclimáticas del sureste de méxico. In J. Martínez Herrera, M. Á. Ramírez Guillermo and J. Cámara-Córdova (Eds.), *Investigaciones Científicas y Agrotecnológicas para la Seguridad Alimentaria* (pp. 181–185).Tabasco, México: Universidad Juárez Autónoma de Tabasco; www.archivos.ujat. mx/2019/div-daca/publicacion/ICASegAlim.pdf.

Carranco, Z. M., Sáenz, E. O., Ornelas, E. G., Barragán, H. B., Ruiz, J. A., Alvarado, R. E. V., et al. (2016). Crecimiento y producción de biomasa de moringa (*Moringa oleifera* Lam.) bajo las condiciones climáticas del Noreste de México. *Tecnociencia Chihuahua*, 10(3): 144–148.

Casanova-Lugo, F., Cetzal-Ix, W., Díaz-Echeverría, V. F., Chay-Canul, A. J., Oros-Ortega, I., Piñeiro-Vázquez, A. T., et al. (2018). *Moringa oleifera* Lam. (Moringaceae): árbol exótico con gran potencial para la ganadería ecológica en el trópico. *Agro Productividad*, 11(2): 100–106.

Castillo-Lopez, R. I., Leon-Felix, J., Angulo-Escalante, M. A., Gutierrez-Dorado, R., Muy-Rangel, M. D. and Heredia, J. B. (2017). Nutritional and phenolic characterization of *Moringa oleifera* leaves grown in Sinaloa, Mexico. *Pakistan Journal of Botany*, 49(1): 161–168.

Castillo-Lopez, R. I., Portillo, L. J. J., León, F. J., Gutiérrez, D. R., Angulo, E. M. A., Muy-Rangel, M. D., and Heredia, J. B. (2018). Inclusion of moringa leaf powder (*Moringa oleifera*) in fodder for feeding Japanese quail (*Coturnix japonica*). *Brazilian Journal of Poultry Science*, 20(1): 15–26.

Dahlgren, J. P., von Zeipel, H., and Ehrlén, J. (2007). Variation in vegetative and flowering phenology in a forest herb caused by environmental heterogeneity. *American Journal of Botany*, 94(9): 1570–1576.

Diniz, F., Pereira, T., Marques, M., and Monteiro, A. (2007). Volatile and non-volatile chemical composition of the white guava fruit (*Psidium guajava*) at different stages of maturity. *Food Chemistry*, 100(1): 15–21.

Duke, J. A. (1978). The quest for tolerant germplasm. In G. Jung (ed.), *Crop Tolerance to Suboptimal Land Conditions. American Society for Agronomy Special Symposium* (Vol. 32, pp. 1–61). Madison, WI: American Society for Agronomy).

Du Toit, E. S., Sithole, J., and Vorster, J. (2020). Pruning intensity influences growth, flower and fruit development of *Moringa oleifera* Lam. under sub-optimal growing conditions in Gauteng, South Africa. *South African Journal of Botany*, 129: 448–456.

Estrada-Hernández, O., Hernández-Rodríguez, O. A., and Guerrero-Prieto, V. M. (2016). Múltiples formas de aprovechar los beneficios de moringa (*Moringa oleifera* Lam.). *TecnoCiencia Chihuahua*, 10(2): 102.

Fahey, J. W. (2005). *Moringa oleifera*: a review of the medical evidence for its nutritional, therapeutic, and prophylactic properties. Part 1. *Trees for Life Journal*, 1(5): 1–15.

Falasca, S. and Bernabé, M. A. (2008). Potenciales usos y delimitación del área de cultivo de *Moringa oleifera* en Argentina. Revista Virtual de REDESMA. http://revistavirt ual. redesma.org/vol3/pdf/investigacion/Moringa.pdf.

Falasca, S. L. and Bernabé, M. A. (2009). Zonificación agroclimática de la Moringa (*Moringa oleifera*) en Argentina para producir biodiesel y bietanol. *Advances en Energías Renovables y Medio Ambiente*, 13, 11–65.

Ferreira, P. M. P., Farias, D. F., Oliveira, J. T. D. A., and Carvalho, A. D. F. U. (2008). *Moringa oleifera*: bioactive compounds and nutritional potential. *Revista de Nutrição*, 21(4): 431–437.

Foidl, N., Mayorga, L., Vásquez, W., Murqueitio, E., Osorio, H., Sanchez, M. D., et al. (1999). Utilización del marango (*Moringa oleifera*) como forraje fresco para ganado. *FAO Animal Production and Health Paper*, 143(1): 341–350.

Foidl, N., Makkar, H. P. S., and Becker, K. (2001). The potential of *Moringa oleifera* for agricultural and industrial uses. In *What development potential for Moringa product?* 20, 45–76.

Fred, F. (1992). *Forestry/Fuel Wood Research and Development Project. Growing Multipurpose Trees on Small Farms*. Bangkok: Winrock International. http://agrss.sherman.hawaii. edu/onfarm/tree/tree0012.

Freiberger, C. E., Vanderjagt D. J., Pastuszyn A., Glew R. S., Mounkaila G., Millson M. et al. (1998). Nutrient content of the edible leaves of seven wild plants from Niger. *Plant Foods for Human Nutrition*, 53(1): 57–69.

Fuglie, L. J. (2001). The miracle tree: The multiple attributes of *Moringa*. Technical Centre for Agricultural and Rural Cooperation (p. 172). Wageningen: Church World Service.

Gandy, M. (2008). Landscapes of disaster: water, modernity, and urban fragmentation in Mumbai. *Environment and Planning A*, 40(1): 108–130.

Garavito, U. (2008). *Moringa oleifera*, alimento ecológico para ganado vacuno, porcino, equino, aves y peces, para alimentación humana, también para producción de etanol y biodiesel, 18. Colombia: Corporación Ecológica Agroganadera SA. www.engormix. com/MA-agricultura/cultivostropicales/articulos/moringa-oleifera-alimento-ecologico-t1891/078-p0.htm.

Gidamis, A. B., Panga, J. T., Sarwatt, S. V., Chove, B. E., and Shayo, N. B. (2003). Nutrient and antinutrient contents in raw and cooked young leaves and immature pods of *Moringa oleifera*, Lam. *Ecology of Food and Nutrition*, 42(6): 399–411.

Godino García, M. (2016). *Moringa oleifera: Árbol multiusos de interés forestal para el sur de la Península Ibérica*. Negocio Agroalimentario y Cooperativo. www.cajamar.es/storage/ documents/020-moringa-v3-1476963334-bf35c.pdf.

Gómez, A. V., and Angulo, K. J. O. (2014). Revisión de las características y usos de la planta *Moringa oleifera*. *Investigación and Desarrollo*, 22(2): 309–330.

Kiragu, J. W., Mathengen, P., and Kireger, E. (2015). Growth performance of *Moringa oleifera* planting materials derived from cuttings and seeds. *International Journal of Plant Sciences and Ecology*, 1(4): 142–148.

Ledea-Rodríguez, J. L., Rosell-Alonso, G., Benítez-Jiménez, D. G., Arias-Pérez, R. C., Ray-Ramírez, J. V., and Reyes-Pérez, J. J. (2018). Producción de semillas de variedades de *Moringa oleifera* Lam en el valle del cauto. *Agronomía Mesoamericana*, 29(2): 415–423.

Mahmood, K. T., Mugal, T., and Haq, I. U. (2010). *Moringa oleifera*: a natural gift – a review. *Journal of Pharmaceutical Sciences and Research*, 2(11): 775.

Makkar, H. P. S., and Becker, K. (1996). Nutritional value and antinutritional components of whole and ethanol extracted *Moringa oleifera* leaves. *Animal Feed Science and Technology*, 63(1–4): 211–228.

Makkar, H. P. S., and Becker, K. (1997). Nutrients and antiquality factors in different morphological parts of the *Moringa oleifera* tree. *Journal of Agricultural Science*, 128(3): 311–322.

Montilla-Mota, J. J., Amundaray, W. G., Gutiérrez, C. E., Fernández-Jiménez, H. G., and Jiménez-Francisco, R. A. (2017). Pre-germination treatments in *Moringa oleifera* seeds and their effect on agronomic variables. *Pastos y Forrajes*, 40(3): 188–194.

Morton, J. F. (1991). The horseradish tree, *Moringa pterygosperma* (Moringaceae) – a boon to arid lands? *Economic Botany*, 45(3): 318–333.

Mota-Fernández, I. F., Valdés-Rodríguez, O. A., and Quintas, G. S. (2019). Caracteristicas socioeconomicas y practicas agricolas de los productores de *Moringa oleifera* Lam. en Mexico. *AGROProductividad*, 12(2): 3–9.

Muñoz, R., Roque, O. G., and Martínez, R. M. (2008). Una solución factible para la clarificación de aguas para consumo humano. Revista Betsime. www. betsime.disaic.cu/secciones/ tec_1_01.htm.

Navarro, S. F. J. (1983). Marco de referencia del área. In: Lépiz, I. R. and Navarro S. F. (Eds.), *Frijol en el noroeste de México Tecnologías de producción*. Mexico: SARH, INIA, CIAPAN, CAEVAC, pp. 1–28.

Nouman, W., Basra, S. M. A., Siddiqui, M. T., Yasmeen, A., Gull, T., and Alcayde, M. A. C. (2014). Potential of *Moringa oleifera* L. as livestock fodder crop: a review. *Turkish Journal of Agriculture and Forestry*, 38(1): 1–14.

Olorukooba, M. M., Mohammed, R., Sodimu, A. I., and Abdullahi, M. U. (2013). Influence of planting methods and pinching on growth and vegetative yield of drumstick (*Moringa oleifera* Lam). *Agrosearch*, 13(1): 143–148.

Olson, M. E. (2001a). Stem and root anatomy of *Moringa* (Moringaceae). *Haseltonia*, 8: 56–96.

Olson, M. E. (2001b). Introduction to the Moringa family. In L. J. Fuglie (ed.), *The Miracle Tree: The Multiple Attributes of Moringa*. Technical Centre for Agricultural and Rural Cooperation (Wageningen: Church World Service): 11–28.

Olson, M. E. (2002). Intergeneric relationships within the Caricaceae-Moringacecae clade (Brassicales) and potential morphological synapomorphies of the clade and its families. International Journal of Plant Sciences 163(1), 51–65.

Olson, M. E., and Alvarado-Cárdenas, L. O. (2016). ¿Dónde cultivar el árbol milagro, *Moringa oleífera*, en México? Un análisis de su distribución potencial. *Revista Mexicana de Biodiversidad*, 87(3): 1089–1102.

Olson, M. E. and Fahey, J. W. (2011). *Moringa oleifera*: a multipurpose tree for the dry topics. *Revista Mexicana de Biodiversidad*, 82(4): 1071–1082.

Olson, M. E., and Razafimandimbison, S. G. (2000). *Moringa hildebrandtii* (Moringaceae): a tree extinct in the wild but preserved by indigenous horticultural practices in Madagascar. *Adansonia*, 22(2): 217–221.

Orduz-Rodríguez, J. O., and Fischer, G. (2007). Balance hídrico e influencia del estrés hídrico en la inducción y desarrollo floral de la mandarina 'Arrayana' en el piedemonte llanero de Colombia. *Agronomía Colombiana*, 25(2): 255–263.

Orwa, C., Mutua, A., Kindt, R., Jamnadass, R., and Anthony, S. (2009). *Moringa oleifera*. Moringaceae. Agroforestry Database: a tree reference and selection guide version. www.worldagroforestry.org/treedb/AFTPDFS/Moringa_oleifera.pdf.

Padilla, C., Fraga, N., and Suárez, M. (2012). Efecto del tiempo de remojo de las semillas de moringa (*Moringa oleifera*) en el comportamiento de la germinación y en indicadores del crecimiento de la planta. *Revista Cubana de Ciencia Agrícola*, 46(4): 419–421.

Padilla, C., Valenciaga, N., Crespo, G., González, D., and Rodríguez, I. (2017). Agronomical requirements of *Moringa oleifera* (Lam.) in livestock systems. *Livestock Research for Rural Development*, 29(11): 4–11.

Palada, M. C. (1996). Moringa (*Moringa oleifera* Lam.): a versatile tree crop with horticultural potential in the subtropical United States. *HortScience*, 31(5): 794–797.

Palada, M. C., and Chang, L. C. (2003). Suggested cultural practices for *Moringa*. Center International Corporation Guide, AVRDC, 03–545.

Paliwal, R., Sharma, V., and Pracheta, J. (2011). A review on horseradish tree (*Moringa oleifera*): a multipurpose tree with high economic and commercial importance. *Asian Journal of Biotechnology*, 3(4): 317–328.

Parrota, J. A. (1993). *Moringa oleifera* Lam: Resedá, Horseradish Tree, Moringaceae, Horseradish-tree Family. International Institute of Tropical Forestry, US Department of Agriculture, Forest Service, 6.

Parrota, J. A. (2004). *Moringa oleifera*. In *Enzyklopädie der Holzgewächse: Handbuch und Atlas der Dendrologie* (pp. 1–8). https://onlinelibrary.wiley.com/doi/abs/10.1002/9783527678518.ehg2005015.

Pérez, A., Sánchez, T., Armengol, N., and Reyes, F. (2010). Características y potencialidades de *Moringa oleifera*, Lamark: Una alternativa para la alimentación animal. *Pastos y Forrajes*, 33(4): 1.

Poehlman, J. M. (1976). *Mejoramiento Genético de las cosechas*. México D.F.: Limusa.

Radovich, T. (2011). Farm and forestry production and marketing profile for Moringa (*Moringa oleifera*). Permanent Agriculture Resources (PAR).

Ramachandran, C., Peter, K. V., and Gopalakrishnan, P. K. (1980). Drumstick (*Moringa oleifera*): a multipurpose Indian vegetable. *Economic Botany*, 34(3): 276–283.

Ramírez, E. R. A., Martínez, J. R. G., Drouaillet, B. E., González, J. C. M., García, A. R. R., and Cancino, S. J. (2020). Variación morfológica en *Moringa oleifera* Lam. a diferentes densidades de población. *Revista Mexicana de Ciencias Agrícolas*, 24: 166–176.

Resmi, D. S., Girija, V. K., and Celine, V. A. (2005). *Fusarium* incited fruit rot of drumstick (*Moringa oleifera* Lamk.). *Journal of Mycology and Plant Pathology*, 35(1): 30–31.

Reyes, S. N. (2006). *Moringa oleifera* and *Cratylia argentea*: potential fodder species for ruminants. Doctoral thesis. Swedish University of Agricultural Sciences, Uppsala, Sweden.

Reyes S. N., and Mendieta Araica, B. (2017). *Guía para el establecimiento y cultivo del marango* (Moringa oleifera) (Managua, Nicaragua: Universidad Nacional Agraria): 40.

Reye,s S. N., Ledin, S., and Ledin, I. (2006). Biomass production and chemical composition of *Moringa oleifera* under different management regimes in Nicaragua. *Agroforestry Systems*, 66(3): 231–242.

Richter, N., Siddhuraju, P. and Becker, K. (2003). Evaluation of nutritional quality of moringa (*Moringa oleifera* Lam.) leaves as an alternative protein source for Nile tilapia (*Oreochromis niloticus* L.). *Aquaculture* 217: 599–611.

Rivero, L. E., Caro, Y., Fernández, L. A., Ayala, L., Rivero, A., and Tamayo, Y. (2020). Comportamiento productivo de cerdos en crecimiento-ceba alimentados con follaje fresco de *Moringa oleifera* Lam., en remplazo parcial de la torta de soya y del maíz. *Livestock Research for Rural Development*, 32(1).

Roloff, A., Weisgerber, H., Lang, U., and Stimm, B. (2009). *Moringa oleifera* LAM., 1785. In *Enzyklopädie der Holzgewächse, Handbuch und Atlas der Dendrologie* (p. 8). www.researchgate.net/profile/John-Parrotta/publication/288327947_Moringa_oleifera/links/5bbe6a67a6fdccf297923720/Moringa-oleifera.pdf.

Saint Sauveur, A. and Broin, M. (2010) Growing and processing Moringa leaves. CDE CTA Moringa Association of Ghana Moringanews. Retrieved from www.anancy.net/documents/file_en/moringawebEN.pdf.

Santoso, B. B., and Parwata, I. A. (2020). The growth of moringa seedling originated from various sizes of stem cutting. *IOP Conference Series: Earth and Environmental Science*, 519(1): 012010.

Schabel, H. G. (2002). Moringa oleifera *Lam. Tropical Tree Seed Manual* (Washington, DC: USDA Forest Service, and Reforestation, Nurseries, and Genetics Resources): 579–581.

Shaykewich, C. F. (1995). An appraisal of cereal crop phenology modeling. *Canadian Journal of Plant Science*, 75(2): 329–341.

Sherry, R. A., Zhou, X. H., and Gu, S. L. (2007). Divergence of reproductive phenology under climate warming. *Proceedings of the National Academy of Sciences of the United States of America*, 104: 198–202.

Tshabalala, T., Ncube, B., Moyo, H. P., Abdel-Rahman, E. M., Mutanga, O., and Ndhlala, A. R. (2020). Predicting the spatial suitability distribution of *Moringa oleifera* cultivation using analytical hierarchical process modeling. *South African Journal of Botany*, 129: 161–168.

Valdés-Rodríguez, O. A., Pérez-Vázquez, A., and Muñoz-García, C. (2018). Efecto de peso y talla de semilla sobre plántulas de *Moringa* y *Ricinus*. *Revista Mexicana de Ciencias Agrícolas*, 9(7): 1411–1422.

Von Maydell, H. J. (1986). *Trees and Shrubs of Sahel, Their Characterization and Uses. Deutsche Gesellschaft fur Technische Zusammenarbeit* (Germany: Eschborn): 334–337.

Wilson, H. K. and Chester, R. A. (1981). *Producción de Cosechas*, 7th edition (p. 411). Mexico: D.F.: CECSA.

Zaku, S. G., Emmanuel, S., Tukur, A. A., and Kabir, A. (2015). *Moringa oleifera*: an underutilized tree in Nigeria with amazing versatility: a review. *African Journal of Food Science*, 9(9): 456–461.

4 Moringa
Phytochemical and Health Benefits

R. Baeza-Jiménez

Laboratorio de Biotecnología y Bioingeniería, Centro de
Investigación en Alimentación y Desarrollo, A.C. Av. Cuarta
Sur 3820, Fracc. Vencedores del Desierto, CP 33089, Delicias,
Chihuahua, México.

R. Flores-Flores

Laboratorio de Biotecnología y Bioingeniería, Centro de
Investigación en Alimentación y Desarrollo, A.C. Av. Cuarta
Sur 3820, Fracc. Vencedores del Desierto, CP 33089, Delicias,
Chihuahua, México.

M.A. Morales-Ovando

Facultad Ciencias de la Nutrición y Alimentos, Universidad
de Ciencias y Artes de Chiapas, Calle Central Norte entre 4ª
y 5ª Norte S/N, CP 30580, Acapetahua, Chiapas, México.

Leticia X. López-Martínez

Laboratorio de Antioxidantes y Alimentos Funcionales,
CONACYT-Centro de Investigación en Alimentación y
Desarrollo, A.C. Carr, Gustavo Enrique Astiazarán Rosas 46,
Col. La Victoria, CP 83304, Hermosillo, Sonora, México.
Corresponding author. E-mail: leticia.lopez@ciad.mx

CONTENTS

DOI: 10.1201/9781003108863-4

4.1 INTRODUCTION

Moringa is considered a traditional crop comprising shrubs and trees, and is extensively cultivated across the tropics: the Pacific islands, the Caribbean, Latin America, Africa, and Asia. Typically, it is consumed as a vegetable, including leaves, roots, and immature pods. For scientific purposes, leaves and seeds have been the most used parts. Leaves are rich in protein, vitamins A, B, C, and minerals, whereas seeds are mainly used for oil extraction (FAO, 2020).

Several studies are reported in technical literature regarding the cultivation, production, and preservation of moringa. Other studies are related to analytical protocols to identify, extract, and characterize the bioactive compounds in *Moringa* due to their beneficial properties, including anti-inflammatory, hypotensive antibiotic, hypocholesterolemic, antitrypanosomal, anti-ulcer, antispasmodic, and hypoglycaemic. In this chapter, the phytochemical compounds and health benefits of moringa are reviewed in chemical composition and bioactive compounds, highlighting the bioactivities exerted for such compounds in different applications in human health and food products.

4.2 CHEMICAL COMPOSITION OF *MORINGA*

Moringa (*Moringa oleifera*) has been reported to be an important source of compounds with antioxidant capacity (FAO/WHO, 1993). Phytochemicals such as β-sitosterol, tocopherols, phenolic acids, carotenoids, glucosinolates, ascorbates, and omega fatty acids have been identified in roots, barks, seeds, and leaves, while phenolic acids, flavonoids, and trace elements that are essential for human health have also been reported in leaves (Ignatov, 2020).

The species of moringa show a chemical composition that includes ash (17.6%), carbohydrates (36%), fiber (9%), lipid (6%), protein (24%), and moisture (7.4%), with an energy supply of 304 kcal/100 g (Ignatov, 2020). Carbohydrates are the main energy source, whereas the ash content indicates that leaves are rich in minerals. Due to its chemical composition (mainly protein content), moringa leaf powders constitute an excellent food source for well-balanced diets for both human and animal nutrition.

4.3 PHYTOCHEMICALS

The presence of important phytochemicals in moringa is responsible for its bioactivities. According to the technical literature, phenolic acids, flavonoids, glucosinolates, isothiocyanates, sterols, and fatty acids have been identified in different parts of moringa, and they have been assayed in various studies. Such phytochemicals are next described.

4.3.1 PHENOLIC ACIDS

Two phenolic acid classes can be differentiated for their structure: derivatives of cinnamic acid (structure C6–C3) and benzoic acid derivatives (structure C6–C1). In general, they comprise of one aromatic ring bound to two propenoic acids (cinnamic acids) or bound to a carboxylic group (benzoic acids) and have different hydroxylation levels (Tomás-Barberán and Clifford, 2000).

Phenolic acids have been mentioned in various studies. The main phenolic acids identified and quantified in moringa are caffeic, chlorogenic, and gallic acids (Valdez-Solana et al., 2015). For example, Singh et al. (2013) noted that gallic acid was the most abundant phenolic acid (1.034 mg/g of dry weight), followed by chlorogenic (from 0.018 to 0.49 mg/g of dry weight) and caffeic acids (0.409 mg/g of dry weight) in dried leaves. In other studies, Rocchetti et al. (2020) indicated the presence of *p*-coumaric acid ethyl ester, syringic acid, gallic acid, gallic acid 4-*O*-glucoside, 3-hydroxybenzoic acid, and feruloylquinic acid with a total content of phenolic acids of 10.60 mg/g of dry weight in methanolic extracts of moringa leaves. Additionally, the presence of gallic acid (0.019 mg/g), *p*-coumaric acid (0.012 mg/g), ferulic acid (0.002 mg/g), caffeic acid (0.038 mg/g), protocatechuic acid (0.0034 mg/g), and cinnamic acid (0.0024 mg/g) have been reported in moringa seed (Singh et al., 2009).

4.3.2 FLAVONOIDS

Flavonoids are a group of compounds (C6–C3–C6), with a general structure of two phenyl rings (A and B) linked through a heterocyclic pyran ring (C). Categorization as flavonols, anthocyanidins, flavanones, flavones, and chalcones is based on the distribution and number of hydroxyl groups, a double bond at position 4 and a double bond between positions 2 and 3 (Middleton et al., 2000). The main flavonoids found in moringa leaves are myricetin (5.8 mg/g), quercetin (0.207 mg/g), and kaempferol (7.57 mg/g; Coppin et al., 2013; Valdez-Solana et al., 2015). Rocchetti et al. (2020) reported the presence of luteolin, luteolin 7-*O*-malonyl-glucoside, luteolin 7-*O*-glucoside, kaempferol 7-*O*-glucoside, quercetin 3-*O*-glucosyl-xyloside, quercetin 3-*O*-xylosyl-rutinoside, and quercetin 3-*O*-rhamnoside in methanolic extracts of *Moringa oliefera* leaves. In a similar study, Atawodi et al. (2010) reported rutin, kaempferol, rhamnoglucoside, quercetin glucoside, and chlorogenic acid. Additionally, Maldini et al. (2014) and Mohammed and Manan (2015) reported rutin, quercetin 3-*O*-β-glucoside, kaempferol 3-*O*-β-glucoside, and kaempferol 3-*O*-rutinoside, as well as a total flavonoid content of 2.9 mg/g in *M. oleifera* seed.

4.3.3 TANNINS

Tannins are considered antinutrients of plant origin, and several studies have mentioned the antioxidant capacity of tannins that do not function only as a primary antioxidant, but also as a secondary antioxidant. Tannins can chelate metal ions such as Fe(II) and inhibit lipid peroxidation by inhibiting cyclooxygenase (Anwar et al., 2007). Some studies have cited the anticancer, antiatherosclerosis and anti-inflammatory properties of tannins (Adedapo et al., 2015). Some researchers described the content of tannins in moringa. In this sense, Teixeira et al. (2014) reported a tannin content range of 13.2–26.3 mg/g in moringa dried leaves. Du Toit et al. (2020) bed described that the tannin content increased across the season in fresh leaves of *M. oleifera* from 24 to 29 mg/g. On the other hand, Hossain et al. (2020) referred to values from 640 to 890 mg/g of tannins in different methanol and ethanol extracts of moringa leaves. Mohammed and Manan (2015) reported values of 0.89 mg/g in total tannin content of moringa seed.

4.3.4 GLUCOSINOLATES

Isothiocyanates are stored as inactive precursors called glucosinolates and can be released through hydrolysis of the enzyme myrosinase after tissue damage. Because mammals, including humans, possess myrosinase-like activity, glucosinolates are also metabolized to isothiocyanates in the digestive tract (Brunelli, 2010).

Glucomoringin is the most abundant glucosinolate in moringa. Maldini et al. (2014) reported values of 28.23, 3.9, and 77.7 mg/100 g in the seed coat, roots, and leaves, respectively, followed by 3-hydroxy-4-(α-L-rhamnopyranosyloxy) benzyl glucosinolate (3.3 mg/100 g) in the seed coat, glucotropaeolin (15.66 mg/100 g) in leaves, glucoraphanin (0.86, 0.58, 2.2 mg/100 g, respectively) in seed coats, roots, and leaves. Other minor components in leaves are 4-(3'-O-acetyl-α-L-rhamnopyranosyloxy) benzyl glucosinolate, 4-(2'-O-acetyl-α-L-rhamnopyranosyloxy) benzyl glucosinolate, 4-(4'-O-acetyl-α-L-rhamnopyranosyloxy) benzyl glucosinolate, and glucoiberin glucosinalbin and in roots and leaves. Glucomoringin content has been reported as >30 mg/g dry weight in seeds and 20 mg/g dry weight in leaves. Glucomoringin acetyl III is present in moringa leaves at levels of 12–17 mg/g dry weight (Amaglo et al., 2010).

4.3.5 STEROLS

The oil extracted from moringa seeds is golden yellow in color and is obtained using solvents (such as *n*-hexane) or by cold pressing, although the latter produces a lower yield (Leone et al., 2016). The composition of sterols in moringa oil comprises β-sitosterol, stigmasterol, campesterol, and D5-avenasterol as the main components, and traces of cholesterol brassicasterol, among others. Anwar et al. (2007) reported that the major sterol components of moringa seed oil are β-sitosterol (46.16%), stigmasterol (18.80%), campesterol (17.59%), and avenasterol (9.26%). Cold press oils obtained from *M. peregrina* and *M. oleifera* seeds contain 23.01 and 36.87 mg/kg β-sitosterol, 11.60 and 12.01 mg/kg stigmasterol, and 6.77 and 8.03 mg/kg campesterol, respectively (Allam et al., 2015). The fraction of sterols is interesting for its key role in cholesterol metabolism by lowering the low-density lipoprotein (LDL) cholesterol in the blood.

4.3.6 FATTY ACIDS

Oleic acid is the main fatty acid in *Moringa* spp seed oils (>74.9%; Anwar et al., 2007); however, Al Juhaimi et al. (2016) reported lower values (51.05% and 7.14%) in *M. molaite* and *M. peregrina* oils, respectively. The major saturated fatty acids are palmitic (12.51%), stearic (2.09%), lauric (1.97%), linolenic acid (1.75%), and linoleic (1.27%). On the other hand, Al Juhaimi et al. (2016) reported that palmitic acid contents of oils ranged from 10.8 in *M. molaite* to 17.06% in *M. peregrina*, stearic acid contents from 1.96 in *M. peregrina* to 2.77% in *M. oleifera*, and linolenic acid from 7.40% (in *M. molaite*) and 32.53% (in *M. peregrina*). In other studies, Compaoré et al. (2011) reported values of 7.8% palmitic acid, 2.2% palmitoleic acid, 7.6% stearic acid, 67.9% oleic acid, and 1.4% linoleic acid in *M. oleifera* seed oil. Janaki (2015) described that *M. oleifera* oil contained 77.40% oleic acid, 12.9% palmitic acid, 2.95% stearic acid, 1.40% margaric acid, 1.40% linoleic acid, and 1.39% α-linolenic acid. Castillo-Lopez et al. (2017) mentioned that in *M. oleifera* leaves, the main fatty acids were linoleic (66.19%), palmitic (17.26%), oleic (2.14%), and stearic (2.66%).

4.4 APPLICATIONS

Moringa is well-known for its content of bioactive compounds that can exert different biological activity, comprising about 40 natural components in leaves, seeds, roots, barks, and flowers (Inbathamizh and Padmini, 2012). Important applications in human health products will now be described (Table 4.1).

TABLE 4.1
Biological activities of different extracts of *M. oleifera*

Type of extract	Bioactive compounds	Bioactivities	References
Aqueous leaf	Quercetin and kaempferol	Antiproliferation and apoptosis cancer cells	Sreelatha et al. (2011)
Hydroethanolic of leaves and seeds	Isothiocyanate, benzyl isothiocyanate and 4-α-rhamnopyranosyloxy) benzyl glucosinolate	Hepatoprotective	Ujah et al. (2013)
Methanolic and ethanolic of leaves and barks	NR	Inhibition of differentiation of colon cancer cells	Lea et al. (2012)
Ethanol of leaves	NR	Upregulation of TNF-α	Akanni et al. (2014)
Methanol of leaves and barks	Flavonoids, terpenoids, glycosides, tannins and saponins	Antioxidant, anti-inflammatory and antinociceptive	Adedapo et al. (2014)
Acetone of leaves and seeds	Phenolic acids, flavonoids	Antimicrobial and antioxidant	Ratshilivha et al. (2014)
Extract of leaves	Isothiocyanates	Antiobesogenic	Waterman et al. (2015)
Ethanol of barks	NR	Antihyperglycemic and hypolipidemic	Irfan et al. (2016)
Aqueous of leaves	Myricetin, apigenin, and chrysin	Immunomodulatory	Kurokawa et al. (2016)
Methanolic of leaves	Gallic, chlorogenic, caffeic, coumaric and ferulic acids	Antioxidant	Castillo-Lopez et al. (2017)

4.4.1 ANTIOXIDANT ACTIVITY

The high content of bioactive compounds, mainly phenolics, in *Moringa* species promotes antioxidant activity. Phenolic compounds neutralized radicals generated in cells by accepting or donating or electrons (Sreelatha and Padma, 2009). The antioxidant capacity of freeze-dried *Moringa* leaves was determined in ethanol and methanol extracts that exhibited 66.8% and 65.1% values for DPPH (1,1-diphenyl-2-picrylhydrazyl) inhibition, respectively. Siddhuraju and Becker (2003) reported that quercetin and kaempferol compounds are responsible for the antioxidant activity. A water extract of *M. stenopetala* leaves exerted higher DPPH inhibition (IC_{50}: 40µg/ml) than a similar extract of *M. oleifera* leaves (IC_{50}: 215µg/ml). Rutin was reported as the major antioxidant compound (Habtemariam, 2015). In the analysis of two variants of *M. oleifera* leaves from Sonora, Mexico, the antioxidant activities reported for oxygen radical absorbance capacity assay was 154.71 and 182.31 µmET/g while the DPPH radical scavenging was 86.82% and 87.92%. A high-performance liquid chromatography analysis showed gallic, chlorogenic, caffeic, coumaric, and ferulic acids (Castillo-López et al., 2017). The aqueous extract of *M. oleifera* leaf extracts significantly reduced DPPH radicals (IC_{50} = 18.5 µg/ml), anion superoxide ($O_2^{\cdot-}$; IC_{50} = 12.7 µg/ml), nitric oxide (NO$^{\cdot}$; IC_{50} = 56.7 µg/ml) and lipid peroxyl radicals (LOO$^{\cdot}$; IC_{50} = 25.3 µg/ml). The results show that extracts of moringa exert potent activity against free radicals that provides protection against oxidative damage and prevents oxidative damage to major biomolecules (Sreelatha and Padma, 2009).

4.4.2 ANTIMICROBIAL POTENTIAL

Rahman et al. (2009) evaluated the antibacterial property of moringa leaf extracts using disc diffusion and the minimum inhibitory concentration (MIC) method against some human pathogenic bacteria and compared to tetracycline as standard. The study disclosed that an ethanol extract of moringa leaves at 1175 µg/disc exhibited antibacterial activity against Gram-negative (*Bacillus cereus*, *B. megaterium*, and *B. subtilis*) and Gram-positive bacteria (*Salmonella shinga*, *S. sonnei*, *Pseudomonas aeruginosa*, and *Pseudomonas* spp.), and the potential antibacterial activity of the extract was similar to the authentic standard tetracycline. The inhibition zones were from 16.25 to 21.5 mm, and MIC values were recorded from 1458 to 916 µg/ml.

Peixoto et al. (2011) evaluated the antimicrobial effect of aqueous (20 g/180 ml) and ethanolic (10 g/190 ml) extracts of moringa leaf by the disc diffusion method at concentrations of 100, 200, 300, and 400 µL against *Escherichia coli*, *Vibrio parahaemolyticus*, *Staphylococcus aureus*, *Aeromonas caviae*, *Enterococcus faecalis*, *P. aeruginosa* and *Salmonella enteritidis*. According to the findings, a concentration of 400 µl of the extracts showed inhibition against *S. aureus*, *A. caviae*, *E. faecalis*, and *V. parahaemolyticus*.

A study conducted by Thilza et al. (2010) evaluated the antimicrobial activity of aqueous extracts by the plate agar diffusion method of moringa leaf against *S. albus*, *P. aerogenosa*, *E. coli*, *S. aureus*, *E. aerogenes*, and *Staphylococcus pyogenus*. At different concentrations (200, 400, 700, and 1000 mg/ml), even at the maximum concentration of the extract, only moderate activity against *Enterobacter aerogenes* was observed. *P. aeruginosa, S. albus, S. aureus*, and *S. pyogenus* were resistant at these concentrations. The highest inhibitory activity was produced at 1000 mg/ml against *E. coli*, less than the standard tetracycline (250 mg/ml).

Saadabi and Zaid (2011) investigated the antimicrobial activity of aqueous extract of moringa seed against *S. aureus, B. subtilis, E. coli*, and *P. aeruginosa*, as well as two species of fungi, *A. niger* and *Candida albicans*, at 5%, 10%, 20%, and 40%. According to MIC, the extracts inhibited the growth of all microorganisms in the range of 11.0–48.0 mg/ml. Moyo et al. (2012) evaluated the activity of an acetone extract of moringa leaves (0.5–5 mg/mL) against *E. coli, Micrococcus kristinae, Enterobacter cloacae, S. aureus*, and *Proteus vulgaris*, determining the MIC and the minimum bactericidal concentration. From their results, an acetone extract at the highest concentration exhibited antibacterial activities against *E. coli, E. cloacae, P. vulgaris, S. aureus*, and *M. kristinae* and at the lowest concentration (0.5 mg/ml) showed bactericidal activity against *E. coli* and *M. kristinae*, and bacteriostatic activity against, *P. vulgaris, S. aureus*, and *E. cloacae*.

Parts of moringa have shown antibacterial activity due to the presence of pterygospermin, 4-α-L-rhamnosyloxy benzyl isothiocyanate, and deoxy-niazimicine (*N*-benzyl, *S*-ethylthioformate; Chaudhary and Chaurasia, 2017). Also, the phenolic compounds that can alter microbial cell membranes generate permeabilization of the cytoplasmic membrane (Paz et al., 2015).

4.4.3 Antiviral Activity

As a traditional medicinal plant, antiviral properties have also been described for moringa species. Niaziminin isolated from moringa leaves showed the capacity to inhibit the Epstein–Barr infection activation in Raji cells (Sudha et al., 2010).

Ethanolic extract of moringa leaves exerted significant antiviral activity against infectious bursal disease virus, rhinovirus, foot and mouth disease virus, and hepatitis virus (Ahmad et al., 2014; Younus et al., 2016). Aqueous extracts of moringa leaf have been shown to reduce the infection rate and diminish herpetic skin lesion increase by activating the cellular immunity in infected mice with herpes simplex virus (HSV-1; Kurokawa et al., 2016). Aqueous moringa seed extract demonstrated potential antiviral capacity against Newcastle disease virus by reducing virus replication (Chollom et al., 2012). Waiyaput et al. (2012) cited that 80% ethanol extract of moringa seeds demonstrated activity against Hepatitis B virus, showing a reduction of the virus replication with moderate cytotoxicity on HepG2 cells. According to Monera and Maponga (2010), extracts from different parts of moringa was likewise used to complement antiretroviral treatment in human immunodeficiency virus (HIV) infection and to treat HIV-related side effects, potentially in early steps in the infection of HIV-1 lentiviral particles in a viral vector-based transmission (Nworu et al., 2013); these antiviral activities may be due to its immunostimulatory effect.

Additionally, it was observed that levels of 25 mg/kg of moringa stimulate the immune system by acting through humoral and cellular immunity in experimental mice models. However, at levels of 750 mg/kg, the high dose was less effective; this behavior could be due to isothiocyanate and glycoside cyanides that act as toxicants generating stress at high concentrations and decreasing the antioxidant potential of moringa (Sudha et al., 2010). According to the studies mentioned above, moringa has potential use in antiviral treatments.

4.4.4 MODULATION OF BLOOD GLUCOSE

Diabetes mellitus (Dm) is described by abnormally high blood glucose levels, either because of insufficient insulin production or its ineffectiveness. There are two clinical types of Dm: type I- and type II-Dm. Type I (TIDm) is insulin-dependent, whereas type II (TIIDm) refers to insulin-resistant diabetes. Dm is linked to obesity with different factors (Wild et al., 2004). Moringa has been mentioned to treat both TIDm and TIIDm, acting as an antidiabetic agent (Tuorkey, 2016). The beneficial activities of moringa leaf on carbohydrate metabolism have been shown by different mechanisms, including improving glucose uptake and utilization, preventing and restoring the integrity and function of β-cells, and increasing insulin activity (Vergara-Jimenez et al., 2017).

Many compounds found in moringa leaves have a positive impact on diabetic pathologies. For example, phenolic acids and flavonoids affect glucose homeostasis function, increase insulin sensitivity in peripheral tissues, and influence β-cell mass tissues (Oboh et al., 2015), and tannins and flavonoids also inhibit intestinal sucrase and pancreatic α-amylase activities (Augustin et al., 2011).

Isothiocyanates have been reported to reduce insulin resistance and hepatic gluconeogenesis (Waterman et al., 2015). Tuorkey (2016) studied the effect of moringa leaf aqueous extract in diabetic rats; they concluded that the extracts decreased insulin resistance levels to combat hyperglycemia. In a similar study, Paula et al. (2016) reported that extracts from moringa leaves improved the immune tolerance in diabetic rats by increasing the proteins CD69 and CD44 and the interferon-gamma (IFN-γ) reduced creatinine and urea levels from damaged kidneys. In addition, the insulin-like protein was observed in the leaves and seed coat of moringa. This protein had antigenic epitopes similar to insulin and showed hypoglycaemic activity on oral administration, reduced blood glucose levels after single and repeated doses of the extract. It displayed antioxidant activity by reducing malondialdehyde and improving catalase levels. It also cross-reacted with anti-insulin antibodies, which proved that moringa extracts have antigenic epitopes like insulin. Moringa leaf extract decreased fasting plasma glucose level, postprandial levels of blood glycated, high-density lipoprotein (HDL)-cholesterol, total cholesterol, very low density lipoprotein (VLDL)-cholesterol, and LDL-cholesterol in type 2 diabetic patients (Kumari, 2010; Nambiar et al., 2010).

Moringa seed extract increased antioxidant enzyme activity and reduced lipid peroxidation in streptozotocin-induced mice (Al-Malki and El Rabey, 2015). It also acts over the reduction of IL-6 that is responsible for glucose homeostasis and pancreatic β-cell activity. The seed extract normalized both the damaged kidney and pancreas activity without altering the organ pathology of treated mice. A study reported that benzylamine isolated from moringa resulted in significant reduction in levels of fasting blood glucose and hyperglycaemic responses of high-fat diet-induced mice (Iffiu-Soltesz et al., 2010).

Barbagallo et al. (2016) described that moringa extracts reduced IL6 expression and had a protective effect on adipocytes by stimulating expression of heme-oxygenase-1 and improved *IRS-1* gene expression (responsible for insulin resistance and a shortage), which are one of the causes of TIIDm. In addition, it stimulated thermogenesis during the differentiation of adipose tissue by upregulating the uncoupling proteins PPARα, sirtuin 1, and coactivator 1 α, which are mediators of thermogenesis. Moringa also demonstrated synergistic effects with sitagliptin, showing delaying lenticular opacity in diabetic rats (Olurishe et al., 2016).

4.4.5 ANTI-INFLAMMATORY ACTIVITY

Roots, seeds, leaves, and extracts from moringa have been used to treat inflammation-related disorders, namely, rheumatoid arthritis, asthma, atopic dermatitis, and allergic rhinitis (Hamza, 2010; Lee et al., 2013). Anti-inflammatory activities could be attributed to 4-[(4'-*O*-acetyl-α-L-rhamnosyloxy) benzyl] isothiocyanate and 4-[(α-L-rhamnosyloxy)-benzyl] isothiocyanate (Stohs and Hartman, 2015); quercetin-3-*O*-glucoside, quercetin, and kaempferol glucosides (Coppin et al., 2013); 4-(3-*O*-acetyl-α-rhamnosyloxy) benzyl isothiocyanate and 4-(2-*O*-acetyl-α-L-rhamnosyloxy) benzyl isothiocyanate (Cheenpracha et al., 2010); 3,5-dihydroxy-6-methyl-2,3-dihydro-4H-pyran-4-1,9-octadecenamide (Inbathamizh and Padmini, 2012); crypto chlorogenic acid (Vongsak et al., 2013); and, 3-dibenzyl urea 4(α-L-rhamnosyloxy)-benzyl glucosinolate (Pandey et al., 2012).

The anti-inflammatory effect of moringa hydroethanolic leaf extracts by assessing NO inhibition production and the expression of pro-inflammatory mediators in macrophages was investigated by Fard et al. (2015). The extracts significantly inhibited the secretion of NO˙ production, interleukin TNF-α, IL-6, IL-1β, and prostaglandin E2. Besides, the bioactive extract induced the production of the anti-inflammatory cytokine IL-10 in a dose-dependent manner. Cheenpracha et al. (2010) have identified NO˙ inhibitors in moringa including phenolic glycosides, 4-[(2'-*O*-acetyl-α-L-rhamnosyloxy) benzyl] isothiocyanate and 4-[(3'-*O*-acetyl-α-L-rhamnosyloxy) benzyl] isothiocyanate. In this sense, Waterman et al. (2015) observed that isothiocyanates isolated from the leaves of *M. oleifera* at a concentration of 1 and 5 µM reduced gene expression and production of inflammatory markers in RAW macrophages, mainly by attenuated expression of inducible nitric oxide synthase, a decrease of production of NO˙, IL-1β, and TNFα. Also, moringa hydroethanolic leaf extract suppressed inflammatory markers' protein expression inducible NO synthase, cyclooxygenase-2 (COX-2), and nuclear factor-kappa (NF-k) in lipopolysaccharide-induced RAW264.7 macrophages in a dose-dependent manner (Alhakmani et al., 2013). Kooltheat et al. (2014) observed that the ethyl acetate extract of *M. oleifera* leaves inhibited the cytokine production of IL-6, IL-8, and TNF-α induced by cigarette smoke with capacities similar to the inhibition of the control drug aspirin. These findings sustain the traditional use of moringa as an effective treatment for inflammation-associated diseases.

4.4.6 ANTICANCER ACTIVITY

Different methods, such as surgery, radiotherapy, and chemotherapy are used to treat cancer, even when they are too expensive and cause several prolonged side effects. Various studies discussed the anticarcinogenic impact of moringa. For example, Jung et al. (2015) suggest that moringa's antiproliferative activity could be to the capacity to scavenge reactive oxygen species in cancer cells, causing cell apoptosis. Table 4.2 summarizes the cancerous cell lines that bioactive compounds in moringa species have inhibited. Kaempferol and quercetin have shown an excellent capacity to reduce hepatocyte growth factor activity by the induction of the expression of the receptor tyrosine kinase MET and phosphorylated MET with IC_{50} values of 6 and 12 µM/l, respectively (Labbe et al., 2009). The chemopreventive capacity of moringa may be attributed to the synergistic action of the bioactive compounds of the extracts, the antioxidant enzymes, and the induction of Phase II enzymes, which may be implicated in the anticarcinogenic activity (Paliwal et al., 2011).

Moringa can be used as an antiproliferative mediator by controlling the development of cancer cells. Moringa extracts exhibited cytotoxic, antileukemia, antimyeloma, and antiproliferative chemoprotective activities (Anwar et al., 2007; Nair and Varalakshmi, 2011) due to the presence of 9-octadecenamide, *cis*-9-hexadecenal, methyl octadecenoate quinic acid, 3,5-dihydroxy-6-methyl-2,3-dihydro-4H-pyran-4-1, quercetin, and kaempferol in flowers (Inbathamizh and Padmini, 2012; Sreelatha

TABLE 4.2
Cell lines studied for anticancer activity of *Moringa* species

Species	Bioactive compounds or extract	Cancerous cell line inhibited	References
M. concanensis	Methanolic and acetone extract of root barks	A549, Hep G2, HT-29	Vijayarajan and Pandian (2016)
M. oleifera	D-Allose, hexadeconoic acid ethyl ester, eugenol and isopropyl isothiocyanate	HCT-8 and MDA-MB-231	Al-Asmari et al. (2015)
	Aqueous leaf extract	A-549	Tiloke et al. (2013)
	Extracts of leaves, 95% ethanol or 50% ethanol	A549, PC-3,T47D, an MCF-7, HCT-16, Colo-205 aTHP-1, HL-60 and K562	Diab et al. (2015)
	Essential oil	HeLa, HepG2, MCF-7, CACO-2 and L929 c	Elsayed et al. (2015)
	Aqueous leaf extract	Human hepatocellular carcinoma HepG2	Jung et al. (2015)
M. peregrina	Ethanolic extracts of the dried leaves	MCF-7, Hep G2, HCT 116	El-Alfy et al. (2011)
	Fatty acids	MCF-7, Hep G2	Abd El Baky and El-Baroty (2013)
M. stenopetala	Benzyl and isobutyl isothiocyanates	HL-60, Hep-G2	Nibret and Wink (2010)

et al., 2011), niazimicin, 4-(4'-*O*-acetyl-α-L-rhamnosyloxy) benzyl isothiocyanate and 4-(L-rhamnosyloxy) benzyl isothiocyanate *O*-ethyl-4-(α-L-rhamnosyloxy) benzyl carbamate, in leaves (Razis et al., 2014), and eugenol and isopropyl isothiocyanate (Khalafalla et al., 2010). Studies show that eugenol has a potent anticancer capability against melanoma, hepatocellular carcinoma, osteosarcoma, and lymphoblastic leukemia (Khalafalla et al., 2010; Manikandan et al., 2011). Sharma et al. (2012) described that the activity of enzymes that protect against carcinogens (glutathione and glutathione *S*-transferase) was recovered by moringa pod extract. Previous studies had also shown the cytotoxicity of different moringa extracts on pancreatic, breast, and colorectal cancers (Lea et al., 2012; Al-Asmari et al., 2015), KB tumor cell lung cancer, alveolar epithelial cancer (Sreelatha et al., 2011; Tiloke et al., 2013; Jung et al., 2015), hepatocellular carcinoma, and melanoma (Khalafalla et al., 2010).

4.4.7 OTHER ACTIVITIES

Antipyretic effects of different moringa seeds extracts (ethanol, petroleum ether, and ethyl acetate) have been observed in rats, using yeast-persuaded hyperpyrexia (Ahmad et al., 2014). Alkaloid moringine acts similarly to ephedrine, relaxing bronchioles so that it could treat asthma. Moringa leaf extract increases wound closure speed and area, demonstrating wound-healing activity (Lambole and Kumar, 2012). Moringa leaf extract improves the restoration of neurotransmitter monoamine levels in the brain, which could be applied to treat Alzheimer's disease (Kirisattayakul et al., 2013).

Vitamin A deficiency is a general cause of vision problems because of decreased dark adaptation to night blindness. Consumption of moringa leaves, leaf powder, or pods rich in vitamin A can prevent eye problems, mainly in children. Ingestion of drumstick leaves with oil can increase vitamin A and delay cataract development (Pullakhandam and Failla, 2007). Additionally, the use of moringa as a high-quality and promissory food was highly accepted for the Integrated Child Development Scheme Supplementary Food Program for its potential as a vitamin A source (Nambiar and Seshadri, 2003).

4.5 CONCLUSION

Moringa is a genus of plants with huge potential in human health and nutrition. According to what has been described in this chapter, moringa is rich in phytochemicals that show several bioactivities. Nevertheless, plenty of traditional uses of moringa leaves and seeds have not been evaluated, especially in species that are not as common as *M. oleifera*. Hence, further research is needed to exploit the different species and many uses. More studies are also required to develop and design functional food based on moringa and moringa compounds.

REFERENCES

Abd El Baky, H.H. and El-Baroty, G.S. (2013). Characterization of Egyptian *Moringa peregrine* seed oil and its bioactivities. *International Journal of Management Sciences and Business Research* 2(7): 98–108.

Adedapo, A., Falayi, O., and Oyagbemi, A. (2014). The antioxidant, anti-inflammatory and antinociceptive activities of the methanol leaf extract of *Moringa oleifera* in some laboratory animals (657.19). *FASEB J* 28(Suppl 1): 657.19.

Adedapo, A.A., Falayi, O.O. and Oyagvemi, A.A. (2015). Evaluation of the analgesic, anti-inflammatory, antioxidant, phytochemical and toxicological properties of the methanolic leaf extract of commercially processed *Moringa oleifera* in some laboratory animals. *Journal of Basic and Clinical Physiology and Pharmacology* 26(5): 491–499.

Ahmad, W., Ejaz, S., Anwar, K. and Ashraf, M. (2014). Exploration of the *in vitro* cytotoxic and antiviral activities of different medicinal plants against Infectious Bursal Disease (IBD) virus. *Central European Journal of Biology* 9(5): 531–542.

Akanni, O.E., Adedeji, A.L. and Oloke, K.J. (2014). Upregulation of TNF-a by ethanol extract of *Moringa oleifera* leaves in benzene-induced leukemic Wister rat: a possible mechanism of anticancer property. *Cancer Research* 74(Suppl 19): 3792–3792.

Al-Asmari, A.K., Albalawi, S.M., Athar, M.T., Khan, A.Q., Al-Shahrani, H. and Islam, M. (2015). *Moringa oleifera* as an anticancer agent against breast and colorectal cancer cell lines. *PLoS ONE* 10: e0135814.

Alhakmani, F., Kumar, S. and Khan, S.A. (2013). Estimation of total phenolic content, in-vitro antioxidant and anti-inflammatory activity of flowers of *Moringa oleifera*. *Asian Pacific Journal of Tropical Biomedicine* 3(8): 623–627.

Al Juhaimi, F., Babiker, E., Ghafoor, K. and Ozcan, M. (2016). Fatty acid composition of three different *Moringa* leave oils. Rivista Italiana Delle Sostanze Grasse 93: 111–113.

Allam, S.M., Aboul-Fotouh, G.E., El-Garhy, G.M., & Gamal, O. (2015). Use of moringa leaves (*Moringa oleifera*) in fattening lambs rations. *Egyptian Journal of Nutrition and Feeds* 18(2 Special): 11–17.

Al-Malki, A.L. and El Rabey, H.A. (2015). The antidiabetic effect of low doses of *Moringa oleifera* Lam. seeds on streptozotocin-induced diabetes and diabetic nephropathy in male rats. *BioMed Research International* Article ID 381040.

Amaglo, N.K., Bennett R.N., Lo Curto, R.B., Rosa, E.A.S., Lo Turco V., Giuffrida A., et al. (2010). Profiling selected phytochemicals and nutrients in different tissues of the multipurpose tree *Moringa oleifera* L., grown in Ghana. *Food Chemistry* 122(4): 1047–1054.

Anwar, F., Latif, S., Ashraf, M., and Gilani, A.H. (2007). *Moringa oleifera*: a food plant with multiple medicinal uses. *Phytotherapy Research* 21(1): 17–25.

Atawodi, S.E., Atawodi, J.C., Idakwo, G.A., Pfundstein, B., Haubner, R., Wurtele, G., et al.(2010). Evaluation of the polyphenol content and antioxidant properties of methanol extracts of the leaves, stem, and root barks of *Moringa oleifera* Lam. *Journal of Medicinal Food* 13: 710–716.

Augustin, J.M., Kuzina, V., Andersen, S.B., and Bak, S. (2011). Molecular activities, biosynthesis and evolution of triterpenoid saponins. *Phytochemistry* 72: 435–457.

Barbagallo, I., Vanella, L., Distefano, A., Nicolosi, D., Maravigna, A., Lazzarino, G., et al. (2016). *Moringa oleifera* Lam. improves lipid metabolism during adipogenic differentiation of human stem cells. *European Review for Medical and Pharmacological Sciences* 20: 5223–5232.

Brunelli, D.T.M. (2010). The isothiocyanate produced from glucomoringin inhibits NF-kB and reduces myeloma growth in nude mice in vivo. *Biochemical Pharmacology* 79(8): 1141–1148.

Castillo-Lopez, R.I., Leon-Felix, J., Angulo-Escalante, M.A., Gutierrez-Dorado, R., Muy-Rangel, M.D., and Heredia, J.B. (2017). Nutritional and phenolic characterization of *Moringa oleifera* leaves grown in Sinaloa, Mexico. *Pakistan Journal of Botany* 49(1): 161–168.

Chaudhary, K. and Chaurasia, S. (2017). Neutraceutical properties of *Moringa oleifera*: a review. *European Journal of Pharmaceutical and Medical Research* 4(4): 646–655.

Cheenpracha, S., Park, E.J., Yoshida, W.Y., Barit, C., Wall, M., Pezzuto, J.M., et al. (2010). Potential anti-inflammatory phenolic glycosides from the medicinal plant *Moringa oleifera* fruits. *Bioorganic & Medicinal Chemistry* 18: 6598–6602.

Chollom, S.C., Agada, G.O.A., Gotep, J.G., Mwankon, S.E., Dus, P.C., Bot, Y.S., et al. (2012). Investigation of aqueous extract of *Moringa oleifera* Lam. seed for antiviral activity against Newcastle disease virus *in ovo*. *Journal of Medicinal Plant Research* 6(22): 3870–3875.

Compaoré, W.R., Nikièma, P.A., Bassolé, H.I.N., Savadogo, A., and Mouecoucou, J. (2011). Chemical composition and antioxidative properties of seeds of *Moringa oleifera* and pulps of *Parkia biglobosa* and *Adansonia digitata* commonly used in food fortification in Burkina Faso. *Current Research Journal of Biological Sciences* 3(1): 64–72.

Coppin, J.P., Xu, Y., Chen, H., Pan, M.H., Ho, C.T., Juliania, R., et al. (2013). Determination of flavonoids by LC/MS and anti-inflammatory activity in *Moringa oleifera*. *Journal of Functional Foods* 5: 1892–1899.

Diab, K.A., Guru, S.K., Bhushan, S. and Saxena, A.K. (2015). In vitro anticancer activities of Anogeissus latifolia, Terminalia bellerica, Acacia catechu and Moringa oleifera Indian plants. *Asian Pacific Journal of Cancer Prevention* 16: 6423–6428.

Du Toit, E.S., Sithole, J., and Vorster, J. (2020). Leaf harvesting severity affects total phenolic and tannin content of fresh and dry leaves of *Moringa oleifera* Lam. trees growing in Gauteng, South Africa. *South African Journal of Botany* 129: 336–340.

El-Alfy, T.S., Ezzat, S.M., Hegazy, A.K., Amer, A.M.M., and Kamel, G.M. (2011). Isolation of biologically active constituents from *Moringa peregrina* (Forssk.) Fiori. (Family: *Moringaceae*) growing in Egypt. *Pharmacognosy Magazine* 7: 109–115.

Elsayed, E.A., Sharaf-Eldin, M.A., and Wadaan, M. (2015). *In vitro* evaluation of cytotoxic activities of essential oil from *Moringa oleifera* seeds on HeLa, HepG2, MCF-7, CACO-2 and L929 cell lines. *Asian Pacific Journal of Cancer Prevention* 16: 4671–4675.

FAO. (2020). www.fao.org/traditional-crops/moringa/en/.

FAO/WHO. (1993). Fats, oils and related products. Food standard program. Codex Alimentarius Commission. Food and Agriculture Organization of the United Nations. World Health Organization, Rome, 8: 33–35.

Fard, M.T., Arulselvan, P., Karthivashan, G., Adam, S.K., and Fakurazi, S. (2015). Bioactive extract from *Moringa oleifera* inhibits the pro-inflammatory mediators in lipopolysaccharide-stimulated macrophages. *Pharmacognosy Magazine* 11(Suppl 4): S556–S563.

Habtemariam, S. (2015). Investigation into the antioxidant and antidiabetic potential of *Moringa stenopetala*: identification of the active principles. *Natural Product Communications* 10(3): 1934578X1501000324.

Hamza, A.A. (2010). Ameliorative effects of *Moringa oleifera* Lam. seed extract on liver fibrosis in rats. *Food Chemistry and Toxicology* 48: 345–355.

Hossain, M.A., Disha, N.K., Shourove, J.H., and Dey, P. (2020). Determination of antioxidant activity and total tannin from drumstick (*Moringa oleifera* Lam.) leaves using different solvent extraction methods. *Turkish Journal of Agriculture-Food Science and Technology* 8(12): 2749–2755.

Iffiu-Soltesz, Z., Wanecq, E., Lomba, A., Portillo, M.P., Pellati, F., Szoko, E., et al. (2010). Chronic benzylamine administration in the drinking water improves glucose tolerance, reduces body weight gain and circulating cholesterol in high-fat diet-fed mice. *Pharmacological Research* 61: 355–363.

Ignatov, I. (2020). Anti-inflammatory and antiviral effects of potassium (K) and chemical composition of *Moringa*. Asian Journal of Biology 9(2): 1–7.

Inbathamizh, L. and Padmini, E. (2012). Gas chromatography–mass spectrometric analyses of methanol extract of *Moringa oleifera* flowers. *International Journal of Chemical and Analytical Science* 2: 1394–1397.

Irfan, H.M., Asmawi, M.Z., Khan, N.A.K. and Sadikun, A. (2016). Effect of ethanolic extract of *Moringa oleifera* Lam. leaves on body weight and hyperglycemia of diabetic rats. *Pakistan Journal of Nutrition* 15: 112.

Janaki, S. (2015). Characterization of cold press *Moringa* oil. *International Journal of Science and Research* 4: 386–389.

Jung, I.L., Lee, J.H., and Kang, S.C. (2015). A potential oral anticancer drug candidate, *Moringa oleifera* leaf extract, induces the apoptosis of human hepatocellular carcinoma cells. *Oncology Letters* 10: 1597–1604.

Khalafalla, M.M., Abdellatef, E., Dafalla, H.M., Nassrallah, A.A., Aboul-Enein, K.M., Lightfoot, D.A., et al. (2010). Active principle from *Moringa oleifera* Lam. leaves effective against two leukemias and a hepatocarcinoma. *African Journal of Biotechnology* 9: 8467–8471.

Kirisattayakul, W., Wattanathorn, J., Tong-Un, T., Muchimapura, S., Wannanon, P. and Jittiwat, J. (2013). Cerebroprotective effect of Moringa oleifera against focal ischemic stroke induced by middle cerebral artery occlusion. Oxidative Medicine and Cellular Longevity 2013: 951415.

Kooltheat, N., Sranujit, R.P., Chumark, P., Potup, P., Laytragoon-Lewin, N. and Usuwanthim, K. (2014). An ethyl acetate fraction of Moringa oleifera Lam. Inhibits human macrophage cytokine production induced by cigarette smoke. Nutrients 6(2): 697–710.

Kumari, D.J. (2010). Hypoglycemic effect of *Moringa oleifera* and *Azadirachta indica* in type-2 diabetes. *Bioscan* 5: 211–214.

Kurokawa, M., Wadhwani, A., Kai, H., Hidaka, M., Yoshida, H., Sugita, C., et al. (2016). Activation of cellular immunity in herpes simplex virus type 1-infected mice by the oral administration of aqueous extract of *Moringa oleifera* Lam. leaves. *Phytotherapy Research* 30: 797–804.

Labbe, D., Provencal, M., Lamy, S., Boivin, D., Gingras, D., and Beliveau, R. (2009). The flavonols quercetin, kaempferol, and myricetin inhibit hepatocyte growth factor-induced medulloblastoma cell migration. *Journal of Nutrition* 139: 646–652.

Lambole, V. and Kumar, U. (2012). Effect of *Moringa oleifera* Lam. on normal and dexamethasone suppressed wound healing. *Asian Pacific Journal of Tropical Biomedicine* 2(1): S219–S223.

Lea, M.A., Akinpelu, T., Amin, R., Yarlagadda, K., Enugala, R., Dayal, B., et al. (2012). Abstract 5439: Inhibition of growth and induction of differentiation of colon cancer cells by extracts from okra (*Abelmoschus esculentus*) and drumstick (*Moringa oleifera*). *Cancer Research* 72(8 Suppl).

Lee, H.J., Jeong, Y.J., Lee, T.S., Park, Y.Y., Chae, W.G., Chung, I.K., et al. (2013). Moringa fruit inhibits LPS induced NO/iNOS expression through suppressing the NFkB activation in RAW264.7 cells. *The American Journal of Chinese Medicine* 41: 1109–1123.

Leone, A., Spada, A., Battezzati, A., Schiraldi, A., Aristil, J., and Bertoli, S. (2016). *Moringa oleifera* seeds and oil: characteristics and uses for human health. Review. *International Journal of Molecular Sciences* 17(12): 2141.

Maldini, M., Maksoud, S.A., Natella, F., Montoro, P., Petretto, G.L., Foddai, M., et al. (2014). *Moringa oleifera*: study of phenolics and glucosinolates by mass spectrometry. *Journal of Mass Spectrometry* 49(9): 900–910.

Manikandan, P., Vinothini, G., Priyadarsini, R.V., Prathiba, D., and Nagini, S. (2011). Eugenol inhibits cell proliferation via NF-κB suppression in a rat model of gastric carcinogenesis induced by MNNG. *Investigational New Drugs* 29(1): 110–117.

Middleton, E., Kandaswami, C., and Theoharides, T.C. (2000). The effects of plant flavonoids on mammalian cells: implications for inflammation, heart disease, and cancer. *Pharmacological Reviews* 52(4): 673–751.

Mohammed, S. and Manan, F.A. (2015). Analysis of total phenolics, tannins and flavonoids from *Moringa oleifera* seed extract. *Journal of Chemical and Pharmaceutical Research* 7(1): 132–135.

Monera, G.T. and Maponga, C.C. (2010). *Moringa oleifera* supplementation by patients on antiretroviral therapy. *Journal of International AIDS Society* 13(4): 188.

Moyo, B., Masika, P.J., and Muchenje, V. (2012). Antimicrobial activities of *Moringa oleifera* Lam. leaf extracts. *African Journal of Biotechnology* 11(11): 2797–2802.

Nair, S. and Varalakshmi, K.N. (2011). Anticancer, cytotoxic potential of *Moringa oleifera* extracts on HeLa cell line. *Journal of Natural Pharmaceuticals* 2: 138–142.

Nambiar, V.S. and Seshadri, S. (2003). Bioavailability trials of beta carotene from fresh and dehydrated drumstick leaves (*Moringa oleifera*) in a rat model. *Plants Foods for Human Nutrition* 56(1): 83–95.

Nambiar, V.S., Guin, P., Parnami, S., and Daniel, M. (2010). Impact of antioxidants from drumstick leaves on the lipid profile of hyperlipidemics. *Journal of Herbal Medicine and Toxicology* 4: 165–172.

Nibret, E. and Wink, M. (2010). Trypanocidal and antileukaemic effects of the essential oils of *Hagenia abyssinica*, *Leonotis ocymifolia*, *Moringa stenopetala*, and their main individual constituents. *Phytomedicine* 17: 911–920.

Nworu, C.S., Okoye, E.L., Ezeifeka, G.O., and Esimone, C.O. (2013). Extracts of *Moringa oleifera* Lam. showing inhibitory activity against early steps in the infectivity of HIV-1 lentiviral particles in a viral vector-based screening. *African Journal of Biotechnology* 12: 4866–4873.

Oboh, G., Agunloye, O.M., Adefegha, S.A., Akinyemi, A.J., and Ademiluyi, A.O. (2015). Caffeic and chlorogenic acids inhibit key enzymes linked to type 2 diabetes (*in vitro*): a comparative study. *Journal of Basic and Clinical Physiology and Pharmacology* 26: 165–170.

Olurishe, C., Kwanashie, H., Zezi, A., Danjuma, N., and Mohammed, B. (2016). Chronic administration of ethanol leaf extract of *Moringa oleifera* Lam. (*Moringaceae*) may compromise glycaemic efficacy of Sitagliptin with no significant effect in retinopathy in a diabetic rat model. *Journal of Ethnopharmacology* 194: 895–903.

Paliwal, R., Sharma, V., Pracheta, Sharma, S., and Yadav, S.N. (2011). Anti-nephrotoxic effect of administration of *Moringa oleifera* Lam. in amelioration of DMBA-induced renal carcinogenesis in Swiss albino mice. *Biology and Medicine* 3(2): 27–35.

Pandey, A., Pandey, R.D., Tripathi, P., Gupta, P.P., Haider, J., Bhatt, S., et al.(2012). *Moringa oleifera* Lam. (Sahijan) – a plant with a plethora of diverse therapeutic benefits: an updated retrospection. *Medicinal and Aromatic Plants* 1: 101–108.

Paula, P.C., Oliveira, J.T., Sousa, D.O., Alves, B.G., Carvalho, A.F., Franco, O.L., et al. (2016). Insulin-like plant proteins as potential innovative drugs to treat diabetes – the *Moringa oleifera* case study. *New Biotechnology* 39: 99–109.

Paz, M., Gúllon, P., Barroso, M.F., Carvalho, A.P., Domingues, V.F., Gomes, A.M., et al. (2015). Brazilian fruit pulps as functional foods and additives: evaluation of bioactive compounds. *Food Chemistry* 172: 462–468.

Peixoto, J.R., Silva, G.C., Costa, R.A., De Sousa Fontenelle, J.R., Vieira, G.H., Filho, A.A., et al. (2011). *In vitro* antibacterial effect of aqueous and ethanolic *Moringa* leaf extracts. *Asian Pacific Journal of Tropical Medicine* 4(3): 201–204.

Pullakhandam, R. and Failla, M.L. (2007). Micellarization and intestinal cell uptake of beta-carotene and lutein from drumstick (*Moringa oleifera*) leaves. *Journal of Medicinal Food* 10: 252–257.

Rahman, M.M., Sheikh, M.M.I., Sharmin, S.A., Islam, M.S., Rahman, M.A., Rahman, M.M., et al. (2009). Antibacterial activity of leaf juice extracts of *Moringa oleifera* Lam. against some human pathogenic bacteria. *Chiang Mai University Journal of Natural Sciences* 8(2): 219–227.

Ratshilivha, N., Awouafack, M.D., du Toit, E.S., and Eloff, J.N. (2014). The variation in antimicrobial and antioxidant activities of acetone leaf extracts of 12 *Moringa oleifera* (Moringaceae) trees enables the selection of trees with additional uses. *South African Journal of Botany* 92: 59–64.

Razis, A.F.A., Ibrahim, M.D., and Kntayya, S.B. (2014). Health benefits of *Moringa oleifera*. *Asian Pacific Journal of Cancer Prevention* 15: 8571–8576.

Rocchetti, G., Pagnossa, J.P., Blasi, F., Cossignani, L., Piccoli, R.H., Zengin, G., et al. (2020). Phenolic profiling and *in vitro* bioactivity of *Moringa oleifera* leaves as affected by different extraction solvents. *Food Research International* 127: 108712.

Saadabi, A.M. and Zaid, I.E.A. (2011). An *in vitro* antimicrobial activity of *Moringa oleifera* L. seed extracts against different groups of microorganisms. *Australian Journal of Basic and Applied Sciences* 5(5): 129–134.

Sharma, V., Paliwal, R., Janmeda, P., and Sharma, S. (2012). Chemopreventive efficacy of *Moringa oleifera* pods against 7,12-dimethylbenz[*a*]anthracene induced hepatic carcinogenesis in mice. *Asian Pacific Journal of Cancer Prevention* 13: 2563–2569.

Siddhuraju, P. and Becker, K. (2003). Antioxidant properties of various solvent extracts of total phenolic constituents from three different agroclimatic origins of drumstick tree (*Moringa oleifera* Lam.) leaves. *Journal of Agricultural and Food Chemistry* 51: 2144–2155.

Singh, B.N., Singh, B.R., Singh, R.L., Prakash, D., Dhakarey, G., Upadhyay, G., et al. (2009). Oxidative DNA damage protective activity, antioxidant and anti-quorum sensing potentials of *Moringa oleifera*. *Food and Chemical Toxicology* 47: 1109–1116.

Singh, R.S.G., Negi, P.S., and Radha, C. (2013). Phenolic composition, antioxidant and antimicrobial activities of free and bound phenolic extracts of *Moringa oleifera* seed flour. *Journal of Functional Foods* 5: 1883–1891.

Sreelatha, S. and Padma, P.R. (2009). Antioxidant activity and total phenolic content of *Moringa oleifera* leaves in two stages of maturity. *Plant Foods for Human Nutrition* 64(4): 303–311.

Sreelatha, S., Jeyachitra, A., and Padma, P.R. (2011). Antiproliferation and induction of apoptosis by *Moringa oleifera* leaf extract on human cancer cells. *Food and Chemical Toxicology* 49: 1270–1275.

Stohs, S.J. and Hartman, M.J. (2015). Review of the safety and efficacy of *Moringa oleifera*. *Phytotherapy Research* 29: 796–804.

Sudha, P., Asdaq, S.M., Dhamingi, S.S., and Chandrakala, G.K. (2010). Immunomodulatory activity of methanolic leaf extract of *Moringa oleifera* in animals. *Indian Journal of Physiology and Pharmacology* 54: 133–140.

Teixeira, E.M.B., Carvalho, M.R.B., Neves, V.A., Silva, M.A., and Arantes-Pereira, L. (2014). Chemical characteristics and fractionation of proteins from *Moringa oleifera* Lam. leaves. *Food Chemistry* 147: 51–54.

Thilza, I.B., Sanni, S., Isah, Z.A., Sanni, F.S., Talle, M., and Joseph, M.B. (2010). *In vitro* antimicrobial activity of water extract of *M. oleifera* leaf stalk on bacteria normally implicated in eye diseases. *Academia Arena* 2(6): 80–82.

Tiloke, C., Phulukdaree, A., and Chuturgoon, A.A. (2013). The antiproliferative effect of *Moringa oleifera* crude aqueous leaf extract on cancerous human alveolar epithelial cells. *BMC Complementary Medicine and Therapies* 13: 226.

Tomás-Barberán, F.A. and Clifford, M.N. (2000). Flavanones, chalcones and dihydrochalcones – nature, occurrence and dietary burden. *Journal of the Science of Food and Agriculture* 80(7): 1073–1080.

Tuorkey, M.J. (2016). Effects of *Moringa oleifera* aqueous leaf extract in alloxan-induced diabetic mice. *Interventional Medicine and Applied Science* 8: 109–117.

Ujah, O.F., Ujah, I.R., Johnson, J.T., Ekam, V.S., and Udenze, E.C.C. (2013). Hepatoprotective property of ethanolic leaf extract of *Moringa oleifera* on carbon tetrachloride (CCl_4) induced hepatotoxicity. *Journal of Natural Products and Plant Resources* 3: 15–22.

Valdez-Solana, M.A., Mejía-García, V.Y., Téllez-Valencia, A., García-Arenas, G., Salas-Pacheco, J., Alba-Romero, J.J., et al.(2015). Nutritional content and elemental and phytochemical analyses of *Moringa oleifera* grown in Mexico. *Journal of Chemistry* Article ID 860381.

Vergara-Jimenez, M., Almatrafi, M.M., and Fernandez, M.L. (2017). Bioactive components in *Moringa oleifera* leaves protect against chronic disease – review. *Antioxidants* 6: 91.

Vijayarajan, M. and Pandian, M.R. (2016). Cytotoxicity of methanol and acetone root bark extracts of *Moringa concanensis* against A549, Hep-G2 and HT-29 cell lines. *Journal of Academia and Industrial Research* 5: 45–49.

Vongsak, B., Gritsanapan, W., Wongkrajang, Y., and Jantan, I. (2013). *In vitro* inhibitory effects of *Moringa oleifera* leaf extract and its major components on chemiluminescence and chemotactic activity of phagocytes. *Natural Product Communications* 8: 1559–1561.

Waiyaput, W., Payungporn, S., Issara-Amphorn, J., and Panjaworayan, T.N. (2012). Inhibitory effects of crude extracts from some edible Thai plants against replication of hepatitis B virus and human liver cancer cells. *BMC Complementary Medicine and Therapies* 12: 246.

Waterman, C., Rojas-Silva, P., Tumer, T.B., Kuhn, P., Richard, A.J., Wicks, S., et al. (2015). Isothiocyanate-rich *Moringa oleifera* extract reduces weight gain, insulin resistance, and hepatic gluconeogenesis in mice. *Molecular Nutrition and Food Research* 59: 1013–1024.

Wild, S., Roglic, G., Green, A., Sicree, R., and King, H. (2004). Global prevalence of diabetes: estimates for 2000 and projections for 2030. *Diabetes Care* 27: 1047.

Younus, I., Siddiq, A., Ishaq, H., Anwer, L., Badar, S., and Ashraf, M. (2016). Evaluation of antiviral activity of plant extracts against foot and mouth disease virus *in vitro*. *Pakistan Journal of Pharmacology Science* 29: 1263–1268.

5 Quality Control and Safety of *Moringa*

Leticia X. López-Martínez
Cátedras CONACYT-Centro de Investigación en Alimentación y Desarrollo, A.C. Carr, Gustavo Enrique Astiazarán Rosas 46, Col. La Victoria, CP 83304, Hermosillo, Sonora, México.

J. Abraham Domínguez-Ávila
Cátedras CONACYT-Centro de Investigación en Alimentación y Desarrollo, A.C. Carr, Gustavo Enrique Astiazarán Rosas 46, Col. La Victoria, CP 83304, Hermosillo, Sonora, México.

Norma Julieta Salazar-Lopez
Coordinación de Alimentos de Origen Vegetal, Centro de Investigación en Alimentación y Desarrollo, A.C., Carr, Gustavo Enrique Astiazarán Rosas 46, Col. La Victoria, Hermosillo, Sonora, 83304, México.

Gustavo A. Gonzalez-Aguilar
Coordinación de Alimentos de Origen Vegetal, Centro de Investigación en Alimentación y Desarrollo, A.C., Carr, Gustavo Enrique Astiazarán Rosas 46, Col. La Victoria, Hermosillo, Sonora, 83304, México.
Corresponding author: e-mail: gustavo@ciad.mx

CONTENTS

DOI: 10.1201/9781003108863-5

5.1 INTRODUCTION

Moringa oleifera Lamarck, or most commonly referred to simply as moringa, is a tree native to India (currently the world's largest producer) and cultivated in other tropical and subtropical regions (Pandey et al., 2011). Although it has been known and widely used for thousands of years, the name moringa has been in use since the eighteenth century. Its nutritional composition is regarded as significant, which has led to moringa being consumed as food in the areas where it is cultivated, with important uses to prevent malnutrition in some Indian and African communities (James and Zikankuba, 2017). Its macro- and micronutrient profile has also been adequate to help satisfy animals' requirements when used to supplement their diets (up to 5%), such as those used to feed poultry, cattle, sheep, pigs, and rabbits (Falowo et al., 2018). Nevertheless, its benefits to human health when consumed as a supplement or herbal medicine have made it increasingly popular with the average consumer, particularly in the west. Most tree tissues are edible and contain a wide variety of phytochemicals. Its leaves and seed pods are of particular interest because they are ground and sold as powder, capsules, dietary supplements, and herbal medicines (Matic et al., 2018). As with other similar products, some claim to exert various specific or vague effects, for example, 'body detoxification', 'energy boost', 'boost skin health', 'prevent signs of aging', 'alkalinize your body', and various others. In fact, some products refer to moringa as a 'superfood' (Saha and Sen, 2019). These products may also be sold in combination with other ingredients or vegetable sources, such as vitamins, minerals, etc. Although there is evidence that moringa consumption does exert some benefits to the consumer, it is important to clearly distinguish between substantiated information, exaggeration, and incorrect affirmations to inform the public better.

According to the many products available, the current market for moringa is extraordinarily successful and owes this success due to many reasons. For example, sedentarism, consumption of processed foods, stress, longer life span, and other factors contribute to a higher incidence of non-communicable diseases, such as obesity, diabetes, and cardiovascular disease (Gonzalez et al., 2017). As the prevalence of these diseases rises, so does the demand for treatments, both allopathic and traditional, including moringa and others. The moringa case is noteworthy because it has been used as a traditional medicine for thousands of years in southern Asia. However, serious scientific evidence has been generated mainly in the past two decades, according to a significant surge in the number of publications and citations on the subject since approximately the year 2000 (Gupta et al., 2020). These publications have described the antioxidant, antiobesity, antidiabetic, anticancer, antiparasitic, antibacterial, and antifungal properties of moringa, among others (Mangundayao and Yasurin, 2017; Ngboula et al., 2018). Thus, its use in traditional medicine, coupled with recently published scientific data, has attracted many new consumers, who aim to use it as a preventive measure or treatment for various non-communicable diseases.

The previously mentioned claims made by moringa-based products and advertisements are possible because they are not sold as medicines, whose labeling is heavily regulated in most jurisdictions, but rather as herbal, dietary, or nutritional supplements, which lack such strict oversight. Traditional medicine developed and flourished across various cultures, relying mainly on locally available plants, with some reports dating back thousands of years and many persisting to this day. Traditional medicine is still the main kind of health care in some populations, in some cases due to insufficient access to modern facilities or simply because of preference for cultural or historical reasons (Redvers and Blondin, 2020). As time passed during the development of traditional medicine, catalogs of plants, tissues, doses, preparation methods, etc., were generated to treat various ailments. Careful recordkeeping was crucial to their success, but the healer's experience was another factor that also contributed because they could modify it as they considered appropriate. Modern medicine still relies on the medical practitioner's judgment, but treatments have been further standardized and targeted for specific applications.

Because traditional medicines were not purified compounds or specifically targeted when first developed, they may have shown varying degrees of effectiveness against multiple diseases, suggesting that a certain treatment was suitable for numerous conditions was common and valid. The modern-day marketing and claims made by moringa-based products (in fact, for any traditional medicine or supplement) is a remnant of these practices, which allow such bold statements. In contrast, allopathic treatments prescribed or administered by medical personnel are based on guidelines backed by specific associations that oversee them and periodically adjust their recommendations according to new evidence (Camacho et al., 2016). Traditional medicines are not subject to the same regulation because they are not prescribed by a medical professional or administered to patients in a hospital setting.

5.2 REGULATION OF *MORINGA* PRODUCTS

Claims that can be made in labeling supplements or traditional medicines have become more restrictive in modern times. However, advertisements still abound where vague or unsupported effects are promised to the consumer, such as those mentioned previously. The claims and language that supplements or herbal medicines can legally use are regulated by national and regional laws and may vary. For example, products sold in the United States must state that 'These statements have not been evaluated by the Food and Drug Administration. This product is not intended to diagnose, treat, cure or prevent any disease', in accordance with 21 CFR § 101 of the United States (www. accessdata.fda.gov/scripts/cdrh/cfdocs/cfcfr/CFRSearch.cfm?fr=101.93).

The European Union, according to Directive 2000/13/EC, is more restrictive, because it states that 'The labeling, presentation, and advertising must not attribute to food supplements the property of preventing, treating or curing a human disease, or refer to such properties' (https://eur-lex.europa.eu/legal-content/EN/TXT/?uri=celex%3A32002L0046). In Mexico, the Federal Commission for the Protection Against Sanitary Risk (COFEPRIS) specifies that the labeling must contain a disclaimer that states that the consumption of this product is the responsibility of whoever recommends it and consumes it, followed by stating that this product is not a medicine, both of which must be stated in bold uppercase letters (www.gob.mx/cofepris/documentos/como-leer-etiquetas-de-los-suplementos-alimenticios).

Because there is evidence in favor of moringa as a health-promoting product, it can be argued that its consumption will exert at least some positive effects. However, there are important aspects that significantly impact its effectiveness. First, the lack of an established therapeutic window means that, although the product may exert a benefit to the consumer, it is possible that the dose consumed is not sufficient or is higher than required to exert it, possibly inducing some adverse effects instead, as documented for various pharmaceuticals (Gillis et al., 2020). Second, imprecise targeting due to the product's many claims further contributes to establishing a therapeutic window; this is best exemplified with drugs whose doses are well-known for different purposes. For example, aspirin is available as 325 mg tablets when intended to be consumed for pain relief, but it is also available as 75 mg tablets when intended to exert an antithrombotic effect. In this case, a 75 mg tablet may not exert effective analgesia, while a 325 mg tablet would supply approximately four times the dose required for an antithrombotic while increasing the risk for adverse side effects (García Rodríguez et al., 2016). Something similar may occur when consuming moringa products for different purposes if the dose is not specified for a particular condition, thereby yielding ineffective results and/or side effects. Third, if therapeutic windows are established for specific applications, standardization must also be guaranteed across multiple manufacturers (Pandey et al., 2016). This would ensure that bioactive compounds' profile is similarly independent of brand, further contributing to effective and consistent results when consuming moringa-based products. This would also allow medical professionals and consumers to avoid potential interactions with other products or treatments, such as those documented for warfarin, where compounds present in some herbal preparations may interfere with this anticoagulant's effectiveness (Smith et al., 2004). Standardizing herbal dosage and preparations would also contribute to the World Health Organization's current plan (2014–2023) to allow people to benefit from both traditional and modern medicine (WHO, 2013).

Thus, there is historical and current evidence that moringa contains macronutrients, micronutrients, and other non-nutritive compounds in most of its tissues. This has led to considerable growth in the number of moringa-based products sold in many international markets, with health claims varying from accurate to vague to incorrect. Lack of precise regulation and standardization makes this a substantial challenge for all traditional medicines, thereby requiring that the average consumer pays attention to the claims made by-products or their advertisements.

5.3 SAFETY OF *MORINGA*

As previously mentioned, moringa and its extracts have been recognized for their potential health benefits due to their phytochemical composition. For example, the beneficial effect of moringa leaves has been associated with the presence of niazimicin A and B, niazirin and niazirinin, 4-[(4'-O-acetyl-alpha-L-rhamnosyloxy) benzyl] isothiocyanate, and polyphenols (antioxidants); quercetin in particular has shown potential for NF-κB-mediated inflammation regulation (Sodvadiya et al., 2020). Seeds are a source of β-sitosterol (with associated antidiabetic and antilipidemic activities), quercetin, chlorogenic acid, and isothiocyanates (antioxidant and anti-inflammatory activity), cationic proteins (anticoagulants), vitamins, micronutrients and minerals, moringinine, moringin isothiocyanates, and glucomoringin (anticarcinogenic, antioxidant, anti-inflammatory, atherosclerotic, antidiabetic and antimicrobial effect; Sodvadiya et al., 2020).

5.3.1 TRIALS IN ANIMALS

The safety of the use of moringa has been evaluated in animals, as shown in Table 5.1. The administration of 1000 mg/kg per day for 90 days of *M. oleifera* leaf powder

TABLE 5.1

Toxicology of *Moringa oleifera* in animals

Source	Animals	Administered via/ Dose	Administration time	Toxicological value	Reference
Aqueous leaf extract of *M. oleifera* Lam	Male Wistar albino mice	Oral (400–6400 mg/kg/) i.p. (250–2000 mg/kg/)	Single dose (evaluated after 24 h of administration)	Oral: No effect observed <6400 mg/kg i.p. LD_{50} 1585 mg/kg	Awodele et al. (2012)
Isothiocyanate enriched hydro-alcoholic extract from *M. oleifera* Lam. seeds	Rats (Sprague–Dawley)	Oral: 78, 257, 772, 2571 mg/kg bw/day (MSE), 30, 100, 300, or 1000 mg/kg bw/day (MIC-1)	14-day repeated-dose	NOAEL: 257 mg/kg/day (MSE) + 100 mg/kg/day (MIC-1)	Kim et al. (2018)
M. oleifera leaf powder	Rats (Sprague–Dawley)	Oral: 2000 mg/kg	Single dose	LD_{50} > 2000 mg/kg	Moodley (2017)
Methanol extract of the seeds of *M. oleifera*	Rats	Oral: 400–5000 mg/kg	–	Median lethal dose: 3870.4 mg/ kg Acute toxicity: 4000 mg/kg Mortality: at 5000 mg/kg No adverse effect at <3000 mg/kg	Ajibade et al. (2014)
M. oleifera aqueous leaf	Rats (male Sprague–Dawley)	Oral: 1000 mg/kg bwt (LD) and 3000 mg/kg bwt (HD)	Single dose	Genotoxic: supra-supplementation levels of 3000 mg/kg LD_{50} ≥3000 mg/kg	Awuku et al. (2012)

Intraperitoneal (i.p.); no observed adverse effect level (NOAEL); moringa isothiocyanate-1 (MIC-1)-enriched hydro-alcoholic; moringa seeds extract (MSE); half-maximal lethal dose (LD_{50}).

(MLP) in mice did not cause differences in the hematological profile and counts of red and white blood cells, glucose, globulin and albumin and, liver enzymes [and aspartate aminotransferase (AST) alanine aminotransferase (ALT)], kidney function (urea, creatinine), cholesterol, or electrolytes (sodium and potassium) (Moodley, 2017). Comparable results were reported by Awuku et al. (2012), who established that the consumption of ≤1000 mg/kg of *M. oleifera* aqueous leaf extract is safe, while no significant changes were observed in hematological values when doses of 1000 and 3000 mg/kg were administered, although, at the highest concentration, reductions in urea and albumin levels were observed, which could be associated with kidney disease.

Aqueous extracts of moringa leaves dispensed orally to rats for 21 days at 2000 mg/kg did not cause mortality. However, doses of 400, 800, and 1600 mg/kg caused variations in hematological parameters (packed cell volume, hemoglobin percentage, total red blood cells, mean corpuscular volume, mean corpuscular hemoglobin concentration, differential and total white blood cells), total proteins and globulins, liver enzymes (ALT, AST, and alkaline phosphatase), bilirubin and urea, without affecting platelet count (Adedapo et al., 2009).

Regarding moringa seed extracts, when 1600 mg/kg of methanolic seed extract was administered to rats, an increase in liver enzymes (AST and ALT) and a decrease in body weight were observed (Ajibade et al., 2014).

Kim et al. (2018) evaluated the safety of moringa isothiocyanate-1 (MIC-1)-enriched moringa seed extract (MSE), with doses of 78–2571 mg/kg/day, and 30–1000 mg/kg/day of MIC-1. The extracts were orally administered for 14 days to rats, and no adverse effect was observed at level with doses ≤2571 mg/kg/day MSE and 100 mg/kg/d MIC-1. However, adverse effects were reported at medium and high doses, including decreased bodyweight, reduced feed intake, irregular respiratory patterns, and piloerection, and enhanced liver weights were observed in females in the mid–high and high dosage groups. At high doses (2571 mg/kg bw/day), mortality, gastrointestinal distention, stomach discoloration, and necrosis of the testicular germinal cells and epididymal cells were detected.

The second most popular moringa species is *M. peregrina*; it is widely distributed from tropical Africa to east India and is considered safe for consumption. El-hak et al. (2018) administered a daily oral dose of 500, 1000, and 2000 mg/kg to albino rats of dry *M. peregrina* seeds for 14 days. No significant changes were observed regarding insulin, albumin, total protein, creatinine, urea, uric acid, follicle-stimulating hormone, luteinizing hormone, and testosterone; however, a significant decrease was noted in serum glucose, cholesterol, triglycerides, and ALT and AST, without histological damage to the liver, pancreas, kidneys, spleen, and testes.

The genotoxic capacity of essential oils from two different varieties of moringa seeds was demonstrated by Elsayed et al. (2019). LD_{50} values of 25.11 and 21.24 µg/ml were observed for *M. peregrina* and *M. oleifera*, respectively. In contrast, concentrations ≥50 µg/ml resulted in 100% mortality of zebrafish embryos. Their anti-angiogenic potential is demonstrated by the fact that 10–20 µg/ml of seed oils interfered with the angiogenic blood vessel development.

5.3.2 HUMAN TRIALS

Daily supplementation with 7 g of MLP during 3 months in postmenopausal women significantly increased ascorbic acid (44.4%), superoxide dismutase (10.4%), glutathione peroxidase (18.0%), hemoglobin content (17.5%), and serum retinol (8.8%), while causing a reduction in malondialdehyde level (16.3%) and fasting blood glucose level (13.5%), which together could induce health benefits (Kushwaha et al., 2014).

The consumption of 100 g of moringa powder weekly by 82 lactating women did not affect their anthropometric values or plasma concentration of ferritin, α-1 acid glycoprotein, or C-reactive protein (Idohou-Dossou et al., 2011). Another study with 140 people with HIV, in which 200 mg of moringa (capsules) were administered, showed that glucose levels were significantly improved while triglycerides remained unchanged. This is consistent with the study carried out by Ahmad et al. (2018), who reported that the consumption of cookies supplemented with moringa decreases postprandial glucose and the sensation of hunger in healthy young subjects. Moringa leaf extracts have also been shown to reduce α-amylase activity by 68.5%, with an $IC_{50} = 220$ µg/ml (Leone et al., 2018).

So far, moringa intake has shown no adverse effects in human studies. Table 5.2 shows clinical studies and interventions with moringa that have been carried out in progress. The health effects of moringa are undoubtedly interesting; however, it is still being studied and requires more clinical studies to establish the benefits and safety of its consumption clearly.

5.4 QUALITY CONTROL OF *MORINGA* PRODUCTS

Most of the marketed herbs are collected from nature; therefore, when monitoring the quality of plant raw materials, particular attention should be given to some uncontrolled conditions of the wild, such as soil type, while country of origin, storing, packing and transportation are also relevant (Sanzini et al., 2011). In general, there is a lack of quality control practices during the preparation and production of moringa products. It is frequently complicated to differentiate between low- and high-quality moringa products, as few official quality standards exist. Currently, quality control of raw plant materials and their products is usually performed using high-performance liquid chromatography (HPLC) and is specific to evaluating one or two indicators. In this sense, Engsuwana et al. (2017) reported a quantitative HPLC assay that could be used for quality control of *M. oleifera* according to phenethyl isothiocyanate identification of astragalin as typical markers in extracts of moringa leaves. Similarly, Sierra-Campos et al. (2020) reported a method using HPLC to identify astragalin as a chemical marker to perform quality control of *M. oleifera* extracts and herbal products, in addition to other characteristic compounds of moringa, such as isoquercetin, syringic acid, chlorogenic acid, rutin, and luteolin.

It is clear that MLP is the most common *M. oleifera* product due to its versatility as an ingredient; however, it is crucial to determine in which category moringa-based products are included. For example, is moringa being consumed as an herb, a tea, or a leafy vegetable? It is also important to emphasize that the permitted levels of

TABLE 5.2
Clinical trials with Moringa (www.clinicaltrials.gov Search Results 10/30/2020)

Title	Conditions	Interventions	Study type/ phase	Outcome measures	Population	Study start/ conclusions
Nutritional impact of *M. oleifera* leaf supplementation in mothers and children	Malnutrition Wasting Growth failure	Supplement: *Moringa oleifera* (high dose) *Moringa oleifera* (low dose) Supplement: Placebo	Interventional	Change in body weight Change in height Head circumference Change in Vitamin A Change in CRP levels Change in fecal neopterin	Enrollment: 500 Age: Child, Adult, Older Adult Sex: All	Study Start: January 2021
Effect of aerobic training and *M. oleifera* on dyslipidemia	Dyslipidemia	Other: Aerobic training and *Moringa oleifera*	Interventional/ not applicable	Positive effect of aerobic training and *M. oleifera* on dyslipidemia and cardiac endurance	Enrollment: 120 Age: 20–50 Sex: Male	February 2020
Effect of *M. oleifera* infusion on health	Metabolic syndrome	Dietary supplement: *Moringa oleifera* tea	Interventional/ not applicable	Change in blood glucose and triglyceride levels Change in low-density lipoprotein	Enrollment: 103 Age: 18–65 Sex: All	February 10, 2020
Effect of moringa capsule in increasing breast milk volume in early postpartum patients	Postpartum women	Drug: *Moringa oleifera* leaf Drug: Placebo	Interventional Phase 4	Breast milk volume at postpartum day 3: Percentage of good satisfaction Quality of life scores Side effects Compliance	Enrollment: 88 Age: 18 and older Sex: Female	July 25, 2020

Study	Conditions	Interventions	Study type / Phase	Outcome measures	Enrollment	Date
Anticariogenic effect of moringa mouthwash compared to chlorhexidine mouthwash	Plaque, dental antimicrobial mouthwash cytotoxicity	Drug: *M. oleifera* Drug: Chlorhexidine mouthwash Drug: Base formula	Interventional Phase 2 Phase 3	Gingival index (GI) White spot lesions	Enrollment: 90 Age: 18–50 Sex: All	November 2020
Moringa leaves for improving hemoglobin, status and underweight among girls in Bangladesh	Impact of moringa leaves on serum hemoglobin among girls	Dietary Supplement: Moringa fry	Interventional Phase 3	Changes in 2 biochemical markers (serum hemoglobin and retinol level) Changes in nutritional status	Enrollment: 226 Age: 12–14 Sex: Female	September 1, 2019
M. oleifera on bone density	Osteoporosis Osteopenia	Dietary supplement: Moringa Cabbage placebo	Interventional/ not applicable	Bone density	Enrollment: 20 Age: 60–70 Sex: Female	July 2015– Completed
Effect of moringa on metformin plasma level in Type 2 Diabetes Mellitus patients	Type 2 Diabetes Mellitus	Dietary supplement: *Moringa oleifera* tea	Interventional Phase: not applicable	Change in fasting blood glucose Change in two-hour post-prandial blood glucose Change in metformin trough plasma concentration Change in metformin peak plasma concentration	Enrollment: 25 Age: 49–77 Sex: All	March 1, 2016 Completed

(continued)

TABLE 5.2 (Continued)
Clinical trials with Moringa (www.clinicaltrials.gov Search Results 10/30/2020)

Title	Conditions	Interventions	Study type/phase	Outcome measures	Population	Study start/conclusions
Moringa delivering nutrition and economic value in Malawi	Malnourishment	Other: Control Corn Soya Diet Other: Test Corn Moringa Diet	Interventional	Change in concentration of phytochemical metabolites in the systemic circulation	Enrollment: 10 Age: 18 years and older Sex: All	September 13, 2018
Effects of moringa on hsCRP and Hgba1c level of patients in hospital in Maynila Medical Center Diabetic Clinic	Diabetes	Dietary supplement: *Moringa oleifera*	Interventional Phase 1	Post-treatments mean HsCRP Post-treatments mean hgba1c	Enrollment: 56 Age: 19–75 Sex: All	August 2014
Cardiovascular and renal effects of moringa extracts and *Stevia rebaudiana* in diabetes mellitus	Capsules in patients with Type 2 Diabetes Mellitus before and after 45 days of add-on therapy	Dietary Supplement: MOROLSTEVER1	Interventional Phase 4	Variation of mitral E' velocity Variation of transmittal flow parameters such as E velocity Urinary excretion of albumin Variation of blood pressure	Enrollment: 19 Age: 21–75 Sex: All	November 1, 2016

pesticides, heavy metals, and microbiological contaminants are often different for fresh or dried herbs, tea, and leafy plants, although they belong to the same plant.

5.4.1 HEAVY METALS

Consumption of plants with a high concentration of heavy metals may cause health problems, significantly impacting renal and hepatic function (Umeh et al., 2020). Heavy metals include essential minerals like iron (Fe), copper (Cu), magnesium (Mg), as well as toxic human elements, such as cadmium (Cd), lead (Pb), mercury (Hg), and arsenic (As), which are toxic to humans even at low concentrations (Ogunfowokan et al., 2019).

Plants take the minerals from the soil, air, and water, where heavy metals are dispersed due to natural and human activities. Concerning the maximum acceptable contamination with toxic heavy metals, there are no worldwide standards available. However, for leafy vegetables that could include moringa, the CAC (Codex Alimentarius Commission) recommends a maximum level for Cd and Pb of 0.3 and 0.2 mg/kg, respectively (CAC, 2015). In the United States, the value of Cd of 0.2 mg/kg was employed by law for fresh herbs and leafy vegetables (EC/178/2006). No specific minimum levels have been specified by EU law for heavy metals in dried herbs. WHO (2005) declared that the maximum permitted concentration of Cd was 0.3 mg/kg, and 10 mg/kg for Pb in leafy vegetables. BPOM RI (National Agency of Drug and Food Republic of Indonesia) declared that the main heavy metal contaminants in leafy vegetables in Indonesia are Pb, Cd, As, and Hg.

A high concentration of heavy metals, exceeding the standard permissible limit of WHO, can cause significant health difficulties (Ogunfowokan et al., 2019). Specific metal identification is necessary for a precise diagnosis due to their toxicity and the possibility of developing syndromes associated with heavy metal poisoning.

The European Pharmacopoeia suggests that inductively coupled plasma (ICP) and atomic absorption spectroscopy (AAS) are suitable methodologies that can quantify heavy metals in raw herbal raw materials and are currently used in the industries.

Sarvesh et al. (2015) analyzed the content of toxic metals to perform quality control of *M. oleifera* bark extracts and identified Pb (<0.010 mg/kg), Hg (<0.156 mg/kg), and Cd (0.002 mg/kg). In another study, Bekoe et al. (2020) evaluated the presence of toxic metals on *M. oleifera* leaves, reporting an absence of As, Pb, and Hg, but with 0.02 mg/kg of Cd. Accordingly, these quality control studies of moringa and its products showed toxic metals values less than the maximum permitted concentration reported by WHO (2005).

5.4.2 PESTICIDES

A major concern for the safety of products from plant origin is pesticide residues used during preharvest or postharvest practices. Methods to prevent excessive concentrations of these compounds are part of acceptable agricultural practices (Schaarschmidt et al., 2016).

Pests and diseases are recognized as the leading cause of the decrease in crop production, thereby resulting in increased use of pesticides to prevent its decrease. However, regular applications and non-controlled use of these chemicals can have unplanned environmental and human health effects. Meeting the minimum requirements of health standards is generally considered one of the most balanced

TABLE 5.3
Limits for some pesticide residues in plant materials

Pesticide	Values (mg/kg)
Alachlor	0.02
Chlordane	0.05
Dichlorodiphenylethanes	1.0
Diazinon	0.5
Dithiocarbamate	2.0
Malathion	0.5
Parathion	1.0

WHO (2007).

agricultural practices. For this, it is necessary to supervise frequent observing of farm produce for pesticide residues and their metabolites to establish if their concentrations meet recommended limits (Table 5.3).

According to their action, pesticides are categorized as fungicides, insecticides, nematicides, herbicides, and others (Britt, 2000). Pesticides can be classified regarding their chemical composition as dichlorodiphenylethanes (DDT), organophosphorus pesticides (OPs; malathion, parathion, and dichlorvos), organochlorine pesticides (OCPs), hexachlorocyclohexanes (HCH), and pesticides of plant origin (pyrethroids and rotenoids). These pesticides and their metabolites are highly fat-soluble, although they are quickly degraded and excreted by humans (WHO, 2007).

OCPs can cause tremors and convulsions, are less toxic than carbamates or organophosphates, remain in the environment, and have a propensity to accumulate in tissue as they move up the food chain, making them unsafe (Britt, 2000).

The most significant adverse consequences related to overexposure to OPs are nervous system signs and symptoms. For example, they inhibit the enzyme acetylcholinesterase, leading to the accumulation of acetylcholine in nervous tissue and at the effector organ, which leads to persistent stimulation of cholinergic synapses. Delayed neuropathy is the most important consequence of chronic exposition to OPs (Britt, 2000). For routine pesticide analysis, the standard methods EN12393 and EN 12396-3 may be used, as well as the WHO (2007) guidelines, which are usually accepted in laboratories expert in contaminant analyses.

Studies related to the pesticide content in moringa and its products are scarce. In this sense, Sarvesh et al. (2015) used a method with quantitative gas chromatography (GC) established by the WHO (2002) in order to perform quality control of *M. oleifera* bark extracts. They identified the presence of chlorinated pesticides (<0.017 mg/kg) and phosphate pesticides (<0.041 mg/kg) at values lower than those established by the WHO guideline.

5.4.3 MICROBIOLOGY

Plants that grow up in a native environment are colonized with microorganisms such as bacteria, yeasts, and molds, meaning that these microorganisms' presence on raw herbals is not necessarily contamination (Dal Molim et al., 2016). Although raw herbal materials should not contain pathogenic species, invasion by some pathogens (Enterobacteriae, Salmonellae, *Escherichia coli*) or potentially pathogenic concentrations of opportunistic species such as *Bacillus* sp. could occur during the post-harvesting manages (Sanzini et al., 2011). Microbiological quality must, therefore, be evaluated regularly in both raw herbal materials and formulations.

Currently, no legal limits have been established for moringa or products derived from it; however, they are comparable concerning raw materials or formulation to other regulated products. For instance, the European Pharmacopoeia analytical methodology (European Pharmacopoeia Commission, and European Directorate for the Quality of Medicines & Healthcare, 2010) is valid. In addition, the microbial limits specified therein are also appropriate for moringa-based supplements; for example, contamination with *Bacillus* sp. should not exceed 10^6 colony forming units (cfu)/g, agreeing to the Health Protection Agency guidelines (HPA, 2004–2009). Regarding *Listeria monocytogenes*, a standardized criterion is for ready-to-eat foods, in which its growth should not occur, specifying amounts over 100 cfu/g as inadequate (Food Standards Australia New Zealand, FSANZ, 2015). The WHO standards (2007) state that the maximum bacterial and fungi count in an herb should be 10^7 and 10^4 cfu/g, respectively. In the *European Pharmacopoeia* (2010), most of the media and diluents used in the routine microbiological examination are specified. They can be applied to diverse materials and products.

In this context, there are few studies related to the microbiological determination of moringa leaves, flowers, seeds, and bark, and most of the existing ones declare the absence of pathogenic microorganisms. Bekoe et al. (2020) evaluated the presence of bacterial and fungi count of the powdered sepals of *M. oleifera*, reporting a bacterial count of 10^7 cfu/g and a fungi count of 10^4 cfu/g. Therefore, these results were within the permissible limit established by the WHO standards (2007).

The effects of drying on the microbiological quality of six different retail of moringa leaves obtained in different markets of Accra (Republic of Ghana) was investigated (Adu-Gyamfi and Mahami, 2014). Samples were dried by mechanical, solar, and room drying. Variation of mean counts of total viable cells, coliforms, molds, and yeasts were 5.92–8.44, 4.85–7.25, 1.65–3.69, and 3.71–4.78 $\log_{10}$ cfu/g, respectively. *E. coli* and *Pseudomonas* spp. showed values of 1.84–4.22 $\log_{10}$ cfu/g, respectively. *Salmonella* spp. was detected in two out of the six samples analyzed. Room-dried leaves showed high counts of total viable cells (6.45), molds and yeasts (3.46), coliforms (6.18), *Pseudomonas* spp. (3.32), and *E. coli* (1.58), when compared to leaves dried by solar or mechanical techniques. The authors concluded that the microbiological quality of most dried moringa leaves sold in Accra does not observe local and international guidelines.

It is necessary to carry out studies on mycotoxins, allergens, dyes, and additives in addition to possible contamination with toxic plants to ensure the quality control of moringa and its products.

5.5 CONCLUSION

The safety and quality control of moringa and its products such as dried leaf, bark, and seeds are essential, whether they are used as a therapeutic therapy or as a dietary complement. Although *M. oleifera* appears to be relatively safe for human consumption, caution should be used when considering its long-term use. Research is also required on the possible mutagenic and carcinogenic effects of *M. oleifera* and its products. For maximum consumer protection, quality control must be applied during the transformation from raw material to finished product. It is clear that no single technique can ensure 100% of the quality of a product; for this, a combination of techniques used in the correct way and utilized to the right production is essential to guarantee the safety and quality of moringa products.

REFERENCES

Adedapo, A.A., Mogbojuri, O.M., and Emikpe, B.O. (2009). Safety evaluations of the aqueous extract of the leaves of *Moringa oleifera* in rats. *Journal of Medicinal Plants Research*, 3: 586–591.

Adu-Gyamfi, A. and Mahami, T. (2014). Effect of drying method and irradiation on the microbiological quality of moringa leaves. *International Journal of Nutrition and Food Science*, 3: 91–96.

Ahmad, J., Khan, I., Johnson, S.K., Alam, I., and Din, Z.U. (2018). Effect of incorporating stevia and moringa in cookies on postprandial glycemia, appetite, palatability, and gastrointestinal well-being. *Journal of the American College of Nutrition*, 37: 133–139.

Ajibade, T., Arowolo, R. and Olayemi, F. (2014). Toxicological evaluation of methanol extract of the seeds of *Moringa oleifera*. *Journal of Pharmacological and Toxicological Methods* 70: 323.

Awodele, O., Adekunle, I., Odoma, S., Teixeira, J.A., and Oluseye, V. (2012). Toxicological evaluation of the aqueous leaf extract of *Moringa oleifera* Lam (Moringaceae). *Journal of Ethnopharmacology*, 139: 330–336.

Awuku, G., Gyan, B., Bugyei, K., Adjei, S., Mahama, R., Addo, P., et al. (2012). Toxicity potentials of the nutraceutical *Moringa oleifera* at supra-supplementation levels. *Journal of Ethnopharmacology*, 139: 265–272.

Bekoe, E.O., Jibira, Y., and Agyei, G.A. (2020). Physicochemical, phytochemical, heavy metal and microbiological analysis of *Moringa oleifera* Lam. leaves. *Pharmacognosy Journal*, 12(6s).

Britt, J. (2000). Properties, and effects of pesticides. In P.L. Williams, R.C. James and S.M. Roberts (eds), *Principles of Toxicology: Environmental and Industrial Applications* (pp. 345–366). New York: John Wiley & Sons Inc.

Camacho, P.M., Petak, S.M., Binkley, N., Clarke, B.L., Harris, S.T., Hurley, D.L., et al. (2016). American Association of Clinical Endocrinologists and American College of Endocrinology Clinical Practice Guidelines for the Diagnosis and Treatment of Postmenopausal Osteoporosis. *Endocrine Practice*, 22(Suppl 4): 1–42.

CAC. Comisión del Codex Alimentarius Límites Máximos de Residuos para Medicamentos Veterinarios en los Alimentos Actualizado en la 35 a Sesión de la Comisón del Codex Alimentarius Commission (2012). Comisión del Codex Alimentarius Límites Máximos del Codex Alimentarius, 40.

Dal Molim, D.G., de Souza Braga, M., Kikuchi, I.S., Brasoveanu, M., Nemtanu, M.R., Dua, K., et al. (2016). The microbial quality aspects and decontamination approaches for the herbal medicinal plants and products: an in-depth review. *Current Pharmaceutical Design*, 22: 4264–4287.

EC Regulation No. 178/2006 of 1 February (2006). *Official Journal of the European Communities L*, 29: 3–24.

El-hak, H.N.G., Raouf, A., Moustafa, A., and Mansour, S.R. (2018). Toxic effect of *Moringa peregrina* seeds on histological and biochemical analyses of adult male albino rats. *Toxicology Reports*, 5: 38–45.

Elsayed, E.A., Farooq, M., Sharaf-eldin, M.A., El-enshasy, H.A.,and Wadaan, M. (2019). Evaluation of developmental toxicity and anti-angiogenic potential of essential oils from *Moringa oleifera* and *Moringa peregrina* seeds in zebra fish (*Danio rerio*) model. *South African Journal of Botany*, 129: 229–237.

Engsuwana, J., Waranuchb, N., Limpeanchobc, N., and Ingkaninana, K. (2017). HPLC methods for quality control of *Moringa oleifera* extract using isothiocyanates and astragalin as bioactive markers. *Science Asia*, 43: 169–174.

European Pharmacopoeia Commission, and European Directorate for the Quality of Medicines & Healthcare. (2010). *European pharmacopoeia*. Vol. 1. Council of Europe, 7th Ed.

Falowo, A.B., Mukumbo, F.E., Idamokoro, E.M., et al. (2018). Multi-functional application of *Moringa oleifera* Lam. in nutrition and animal food products: a review. *Food Research International*, 106: 317–334.

Food Standards Australia New Zealand. (2015). About FSANZ. www.foodstandards.gov.au/about/Pages/default.aspx (accessed October 2019).

García Rodríguez, L.A., Martín-Pérez, M., Hennekens, C.H., Rothwell, P.M., and Lanas, A. (2016) Bleeding risk with long-term low-dose aspirin: a systematic review of observational studies. *PLoS One*, 11(8): e0160046.

Gillis, A., Sreenivasan, V., and Christie, M.J. (2020). Intrinsic efficacy of opioid ligands and its importance for apparent bias, operational analysis, and therapeutic window. *Molecular Pharmacology*, 98: 410–424.

Gonzalez, K., Fuentes, J., and Marquez, J.L. (2017). Physical inactivity, sedentary behavior and chronic diseases. *Korean Journal of Family Medicine*, 38(3): 111–115.

Gupta, B.M., Singh, S. and Verma, D. (2020). *Moringa oleifera*: a bibliometric analysis of international publications during 1935–2019. *Pharmacognosy Reviews*, 14(28): 82–90.

Health Protection Agency. (2004). Health Protection Agency – Corporate Plan 2004–2009. Available from: www.hpa.org.uk/webc/HPAwebFile/HPAweb_C/1197021714519.

Idohou-Dossou, N., Diouf, A., Gueye, A., Guiro, A., and Wade, S. (2011). Impact of daily consumption of moringa (*Moringa oleifera*) dry leaf powder on iron status of Senegalese lactating women. *African Journal of Food Agriculture Nutrition and Development*, 11: 4985–4999.

James, A. and Zikankuba, V. (2017). *Moringa oleifera* a potential tree for nutrition security in sub-Sahara Africa. *American Journal of Research Communication*, 5(4): 1–14.

Kim, Y., Jaja-Chimedza, A., Merrill, D., Mendes, O.M., and Raskin, I. (2018). A 14-day repeated-dose oral toxicological evaluation of an isothiocyanate- enriched hydro-alcoholic extract from *Moringa oleifera* Lam. seeds in rats. *Toxicology Reports*, 5: 418–426.

Kushwaha, S., Chawla, P., and Kochhar, A. (2014). Effect of Supplementation of drumstick (*Moringa oleifera*) and amaranth (*Amaranthus tricolor*) leaves powder on antioxidant profile and oxidative status among postmenopausal women. *Journal of Food Science and Technology* 51(11): 3464–3469.

Leone, A., Bertoli, S., Di Lello, S., Bassoli, A., Ravasenghi, S., Borgonova, G., et al. (2018). Effect of *Moringa oleifera* leaf powder on postprandial blood glucose response: *in vivo* study on saharawi people living in refugee camps. *Nutrients*, 10: 1494.

Mangundayao, K. and Yasurin, P. (2017). Bioactivity of *Moringa oleifera* and its applications: a review. *Journal of Pure Applied Microbiology*, 11: 43–50.

Matic, I., Guidi, A., Kenzo, M., Mattei, M., and Galgani, A. (2018). Investigation of medicinal plants traditionally used as dietary supplements: a review on *Moringa oleifera*. *Journal of Public Health in Africa*, 9: 841–841.

Moodley, I. (2017). Acute toxicity of *Moringa oleifera* leaf powder in rats. *Journal of Medicinal Plants Studies*, 5: 180–185.

Ngboula, K., Liyongo, C.I., Bongo, G.N., Ashande, C.M, Lufualobo, L.G., Gbolo, B.Z., et al. (2018). An updated review on the bioactivities and phytochemistry of the nutraceutical plant *Moringa oleifera* Lam (Moringaceae) as valuable phytomedicine of multipurpose. *Discovery Phytomedicine*, 5(4): 52–63.

Ogunfowokan, A.O., Adekunle, A.S., Oyebode, B.A., Oyekunle, J., Komolafe, A., and Omoniji-Esan, G. (2019). Determination of heavy metals in urine of patients and tissue of corpses by atomic absorption spectroscopy. *Chemistry Africa*, 699–712.

Pandey, A., Pradheep, K., Gupta, R., Nayar, E.R., and Bhandari, D.C. (2011). Drumstick tree (*Moringa oleifera* Lam.): a multipurpose potential species in India. *Genetic Resources and Crop Evolution*, 58: 453–460.

Pandey, R., Tiwari, R.K., and Shukla, S.S. (2016). Omics: a newer technique in herb. Drug standardization and quantification. *Journal of Young Pharmacists*, 8(2): 76–81.

Redvers, N., and Blondin, B. (2020). Traditional indigenous medicine in North America: a scoping review. *PLoS ONE*, 15(8): e0237531.

Saha, G. and Sen, M. (2019). Moringa production and consumption: an alternative perspective for government policy-making. In A.K.J.R. Nayak (ed.), *Transition Strategies for Sustainable Community Systems* (pp. 183–194). Cham: Springer.

Sanzini, E., Badea, M., Dos Santos, A., Restani, P., and Sievers, H. (2011). Quality control of plant food supplements. *Food and Function*, 2: 740–746.

Schaarschmidt, S. (2016). Public and private standards for dried culinary herbs and spices – part I: standards defining the physical and chemical product quality and safety. *Food Control*, 70: 339–349.

Sarvesh, K., Kumar, M.S., Ankit, S., and Kumar, S.A. (2015). Quality control standardization of the bark of *Moringa oleifera* Lam. *International Journal for Pharmacy and Pharmaceutical Science*, 7: 56–60.

Sierra-Campos, E., Valdez-Soana, M.A., Avitia-Domínguez, C.I., Téllez-Valencia, A., Meza-Velázquez, J.A., Aguilera-Ortíz, M., et al. (2020). Standardization based on chemical markers of *Moringa oleifera* herbal products using bioautography assay, thin layer chromatography and high-performance liquid chromatography-diode array detector. *Malaysian Journal of Analytical Sciences*, 24: 449–463.

Smith, L., Ernst, E., Ewings, P., Myers, P., and Smith, C. (2004). Co-ingestion of herbal medicines and warfarin. *British Journal of General Practice*, 54: 439–441.

Sodvadiya, M., Patel, H., Mishra, A., and Nair, S. (2020). Emerging insights into anticancer chemopreventive activities of nutraceutical *Moringa oleifera*: molecular mechanisms, signal transduction and *in vivo* efficacy. *Current Pharmacology Reports*, 6: 38–51.

Umeh, C., Asegbeloyin, J.N., Akpomie, K.G., Oyeka, E.E., and Ochonogor, A.E. (2020). Adsorption properties of tropical soils from Awka North Anambra Nigeria for lead and cadmium ions from aqueous media. *Chemistry Africa*, 3: 199–210.

World Health Organization. (2002). WHO Quality Control Methods for Medicinal Plant Materials. Geneva: World Health Organization.

World Health Organization. (2005). *National Policy on Traditional Medicine and Regulation of Herbal Medicines: Report of a WHO Global Survey*. Geneva: World Health Organization.

World Health Organization. (2007). *WHO Guidelines for Assessing Quality of Herbal Medicines with Reference to Contaminants and Residues*. Geneva: World Health Organization.

World Health Organization. (2013). *WHO Traditional Medicine Strategy: 2014–2023*. Geneva: World Health Organization.

6 Phytochemicals from *Moringa* Species

Luis A. Cabanillas-Bojórquez
Centro de Investigación en Alimentación y Desarrollo A.C., Carretera a Eldorado Km. 5.5, Col Campo El Diez, Culiacán, Sinaloa, 80110 México.

Erick Paul Gutiérrez-Grijalva
Cátedras CONACYT-Centro de Investigación en Alimentación y Desarrollo A.C., Carretera a Eldorado Km. 5.5, Col Campo El Diez, Culiacán, Sinaloa, 80110 México.

Melissa García-Carrasco
Centro de Investigación en Alimentación y Desarrollo A.C., Carretera a Eldorado Km. 5.5, Col Campo El Diez, Culiacán, Sinaloa, 80110 México.

J. Basilio Heredia
Centro de Investigación en Alimentación y Desarrollo A.C., Carretera a Eldorado Km. 5.5, Col Campo El Diez, Culiacán, Sinaloa, 80110 México.
Corresponding author: jbheredia@ciad.mx

CONTENTS

6.1 INTRODUCTION

Plants of the genus *Moringa* belong to the Moringaceae family. It is generally accepted that the *Moringa* genus comprises around 13 species distributed worldwide in southwest Asia, southwest and northeast Africa, Madagascar, and Northwestern India (Abd Rani et al. 2018; Vergara-Jimenez et al. 2017). The name "moringa" is possibly derived from the Tamil word 'murunggi' or the Malayalam word 'muringa' (Senthilkumar et al. 2018). Plants of this genus have been used for thousands of years, and there is a record of their multiple uses in ancient Indian, Greek, and Egyptian civilizations (Senthilkumar et al. 2018).

The *Moringa* genus is characterized by its resistance to adverse climate conditions, such as drought and frost. The most-reported *Moringa* species are *M. oleifera*, *M. stenopetala*, *M. concanensis*, and *M. peregrina*; the other species are endemic to African countries and Madagascar, which limits their study. Furthermore, *Moringa* species can be classified into three groups based on their type of stem (Vergara-Jimenez et al. 2017).

- Bottle trees: *Moringa* species of this type have bloated stems to store water; *M. stenopetala*, *M. drouhardii*, *M. ovalifolia*, and *M. hildebrandtii* are considered within this classification.
- Slender trunks: *M. peregrina*, *M. concanensis*, and *M. oleifera*.
- Tuberous shrubs: *M. arborea* Verdcourt, *M. borziana* Mattei, *M. longituba* Engler, *M. pygmaea* Verdcourt, *M. rivae* Chiovenda, and *M. ruspoliana* Engler.

However, *M. oleifera* is by far the most evaluated and reported of all the species. This species has been used as food, medicine, cosmetic oil, or animal feed (Vergara-Jimenez et al. 2017). On the nutritional aspect, *M. oleifera* is valued for its nutritional properties as a source of proteins, dietary fiber, and minerals (Dhakad et al. 2019); for its economic importance, its medicinal and nutraceutical potential (Gupta et al. 2018). *Moringa* species have been used in folk medicine as a remedy against diabetes,

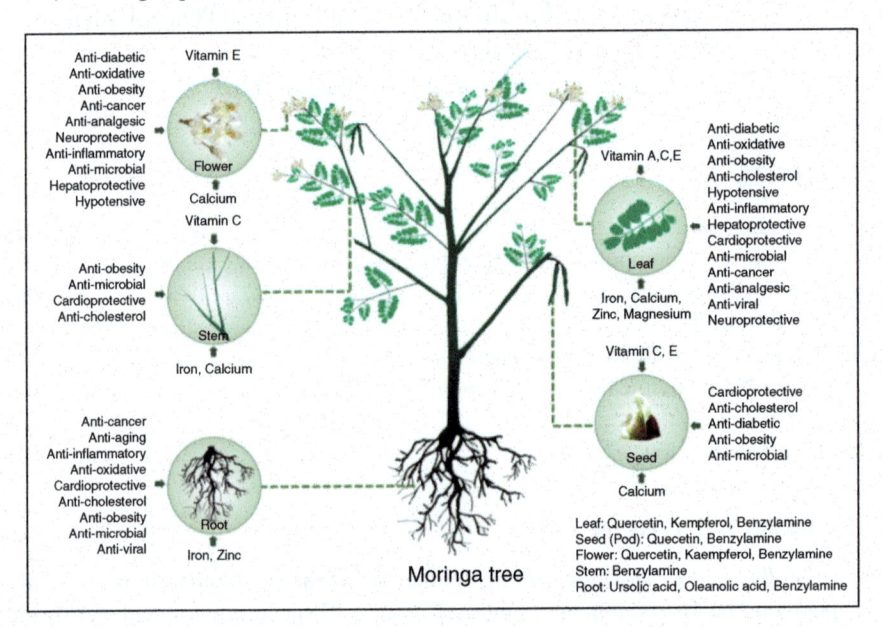

FIGURE 6.1 Functional properties of different vegetative parts of a *Moringa* tree.

wound healing, fever, and hypertension, among other ailments (Senthilkumar et al. 2018). More information about the functional properties of the *Moringa* trees can be observed in Figure 6.1.

Recent studies have shown that *Moringa* species are a rich source of phytochemicals; these molecules are biosynthesized as a defense mechanism against biotic and abiotic stress conditions. Moreover, these molecules have been widely associated with the ethnopharmacological potential of medicinal plants traditionally used in folk medicine in different cultures worldwide. This chapter comprises recent studies on the identification of phytochemicals in *Moringa* species and with ethnopharmacological potential.

6.2 POLYPHENOL COMPOUNDS FROM *MORINGA* SPECIES

Moringa species are rich sources of phytochemicals such as polyphenol compounds, which have been found in the different parts of the plant, such as flowers, husk,

TABLE 6.1
Polyphenol compounds in *Moringa* species

Species	Plant part	Compounds	Identification technique	Reference
Moringa citrifolia	Leaves	Quercetin, kaempferol, chlorogenic acid, and ferulic acid	HPLC	Andarwulan et al. (2012)
Moringa oleifera	Leaves	4-*O*-caffeoylquinic acid, 3-caffeoylquinic acid, rutin, quinic acid, quercitrin, caffeic acid, quercetin, kaempferol, apigenin, luteolin, cirsiliol, naringenin, benzoic acid, ferulic acid, chlorogenic acid, sinapic acid, 3-caffeoylquinic acid, coumaroylquinic acid, caffeoylquinic acid isomers, apigenin glucoside isomers, quercetin-acetyl-glycoside, quercetin 3-*O*-glucoside, kaempferol-3-*O*-glucoside, kaempferol-acetyl-glycoside isomer, and cirsilineol	HPLC	Oldoni et al. (2019); Xu et al. (2019); Zhao et al. (2019); Bennour et al. (2020); Ibrahim and Ghareeb (2020); Saleem et al. (2020)
	Seed	Gallic acid, ellagic acid, kaempferol, *p*-hydroxybenzoic acid, and quercetin	UFLC-MS	Chhikara et al. (2020); Fejer et al. (2019)

(*continued*)

TABLE 6.1 (Continued)
Polyphenol compounds in *Moringa* species

Species	Plant part	Compounds	Identification technique	Reference
Moringa ovalifolia	Leaves	Kaempferol rutinoside, quercetin rutinoside, 3-acyl, 4-acyl, 5-acyl *p*-coumaroylquinic acid, caffeoylquinic acid, feruloylquinic acid, and isorhamnetin rutinoside	UHPLC-MS, UPLC-MS	Makita et al. (2016); Makita et al. (2017)
Moringa peregrina	Aerial parts	Quercetin, rutin, chrysoeriol-7-*O*-rhamnoside, and 6,8,3',5'-tetramethoxy apigenin	NMR, TLC	Elbatran et al. (2005)
	Leaves	Rutin, gallic acid, protocatechuic acid, catechin, hydroxybenzoic acid, caffeic acid, syringic acid, chlorogenic acid, fisetin, quercetin, naringenin, and isorhamnetin	HPLC, LC-MS	Dehshahri et al. (2012); Al Juhaimi et al. (2017)
Moringa pterygosperma	Leaves	Chlorogenic acid, caffeic acid, luteolin, quercetin, kaempferol and ferulic acid	HPLC	Andarwulan et al. (2012)

HPLC = high-performance liquid chromatography; UFLC = ultra-fast liquid chromatography; MS = mass spectrometry; UPLC = ultra-performance liquid chromatography; NMR = nuclear magnetic resonance; TLC = thin layer chromatography; LC = liquid chromatography.

leaves roots, and seeds (Table 6.1 and Figure 6.1; Chhikara et al. 2020; Elbatran et al. 2005; Fejer et al. 2019; Guzmán-Maldonado et al. 2020; Saleem et al. 2020). *Moringa oleifera* is the most studied species due to a high polyphenolic content in different plant parts (Fejer et al. 2019; Saleem et al. 2020; Xu et al. 2019; Zhao et al. 2019).

- Flowers. The polyphenol content of *M. oleifera* flowers has been studied; it has been reported that flowers from Mexico have a total phenolic content of 1908.71 ± 110.74 mg gallic acid equivalent/100 g and a total flavonoid content of 936.44 ± 88.03 mg gallic acid equivalent/100 g (Guzmán-Maldonado et al. 2020).
- Husk. Guzmán-Maldonado et al. (2020) also reported a total phenolic content of 3.65–4.5 mg gallic acid equivalent/100 g and a total flavonoid content of 202.81–1427.27 mg gallic acid equivalent/100 g from Mexican *M. oleifera* husk.

- Leaves. *Moringa* leaves are an important source of polyphenolic compounds. Saleem et al. (2020) identified polyphenolic compounds by high-performance liquid chromatography (HPLC) and reported that different solvent extracts from *M. oleifera* leaves have a total phenolic content of 71.19–91.37 mg gallic acid equivalent/g extract and total flavonoid content of 53.28–65.7 mg catechin/g extract.

Similarly, Ibrahim and Ghareeb (2020) found flavonoids and phenolic acid in methanolic extracts from *M. oleifera* leaves and reported a total phenolic content of 210.75 ± 4.95 mg gallic acid equivalent/g extract. Also, Moyo et al. (2012) stated a phenolics, flavonoids, flavonols, and proanthocyanidins content of 120.33 ± 0.76 tannic acid equivalent/g, 295.01 ± 1.99 quercetin equivalent/g, 132.74 ± 0.50 quercetin equivalent/g, and 32.59 ± 0.5 catechin equivalent/g, respectively, from acetonic *M. oleifera* leaf extracts.

Likewise, it has been reported that the phytochemical content from plants such as *Moringa* varied with the environment, so *Moringa* leaves from different places have been studied (Bennour et al. 2020; Fejer et al. 2019; Guzmán-Maldonado et al. 2020; Xu et al. 2019; Zhao et al. 2019). In this sense, two *M. oleifera* leaves from Mexico have a total phenolic content of 20.25–26.14 mg gallic acid equivalent/100 g and total flavonoids of 1427.27–2859.44 mg gallic acid equivalent/100 g (Guzmán-Maldonado et al. 2020). In addition, Bennour et al. (2020) studied the phytochemical composition of *M. oleifera* leaves from southern Tunisia; they found that a methanolic extract has a total phenolic content of 80.9–136.44 mg gallic acid equivalent/g extract, as well as total flavonoid content of 31.7–44.2 mg quercetin equivalent/g extract. Likewise, it has been reported that *n*-butanol was the best solvent for flavonoid extraction (1429.34 mg rutin equivalent/g) from Chinese *M. oleifera* leaves (Zhao et al. 2019). Similarly, Kenyan *M. oleifera* leaves were studied by Xu et al. (2019), who found a total flavonoid content of 192.36 ± 2.96 mg rutin equivalent/g. On the other hand, Caribbean *M. oleifera* leaves have a total phenolic content of 23.7–635.6 mg gallic acid equivalent/g (Fejer et al. 2019).

Other *Moringa* species leaves have also been studied; for instance, Dehshahri et al. (2012) reported a total phenolic content of 88.05 ± 1.08 mg gallic acid equivalent/g from *Moringa peregrina* leaves. Similarly, Al-Owaisi et al. (2014) studied different solvent extractions of *M. peregrina* leaves. They found that methanolic extract has a higher total phenolic and flavonoid extraction (94.56 ± 3.53 mg gallic acid equivalent/g and 20.84 ± 4.02 mg quercetin equivalent/g, respectively). Ullah et al. (2015) and Al Ameri et al. (2014) reported that the hydroalcoholic and ethanolic extracts of *M. peregrina* leaves have phytochemicals such as total phenolic compounds (11% and 0.2 mg tannic acid equivalent/ml, respectively). Al Juhaimi et al. (2017) found that methanolic extracts of young leaves *M. peregrina* and *M. oleifera* have anthocyanin, total flavonoid, and total phenolic content of 22.90 ± 1.70–23.10 ± 1.30 mg cyanidin-3-glucoside/g, 35.5 ± 3.4–38.90 ± 1.90 mg catechol equivalent/g, and 45.9 ± 0.4–56.10 ± 0.90 mg gallic acid equivalent/g, respectively. On the other hand, Toma et al. (2014) reported that a hydroalcoholic extract of *M. stenopetala* leaves has flavonoid and phenolic compound content of 71.73 ± 2.48 mg quercetin equivalent/g crude extract and 79.81 ± 2.85 mg gallic acid equivalent/g.

6.2.1 ROOTS

There are few available reports of phytochemical compounds from *Moringa* roots. One of the few studies was published by Xu et al. (2019). They reported that Kenyan *M. oleifera* roots are a flavonoid source with a total flavonoid content of 106.79 ± 2.12 mg rutin equivalent/g.

6.2.2 SEEDS

Özcan (2020) reviewed the phytochemical composition of seed in *Moringa* species: *M. oleifera* seed contains flavonoids, phenolic acids, and proanthocyanidins. Guzmán-Maldonado et al. (2020) reported that *M. oleifera* seeds have a total phenolic content of 160.87–184.05 mg gallic acid equivalent/100 g and a total flavonoid content of 149.47–249.86 mg gallic acid equivalent/100 g. Likewise, it has been reported that *M. oleifera* seeds have a total phenolic content of 45.81 mg gallic acid equivalent/g and a total flavonoid content of 9.93 mg quercetin equivalent/g (Chhikara et al. 2020). Similarly, Anunciacao et al. (2020) reported that *M. oleifera* ethanolic seed extract has a total phenolic content of 235.11 ± 73.49 mg gallic acid equivalent/g extract. Caribbean *M. oleifera* seeds have a total phenolic content of 229.5 ± 34.57 mg gallic acid equivalent/g (Fejer et al. 2019). On the other hand, the total flavonoid content of 5.89 ± 0.65 mg rutin equivalent/g from Kenyan *M. oleifera* seeds was reported by Xu et al. (2019).

6.3 CAROTENOIDS FROM *MORINGA* SPECIES

Moringa citrifolia, M. oleifera, and *M. pterygosperma* are sources of carotenoids such as lutein, β-carotene, and zeaxanthin, among others (Saini, Harish Prashanth, et al. 2014; Saini, Shetty, and Giridhar 2014; Saini, Shetty, et al. 2014). Likewise, the carotenoid content of different plant parts has been identified and summarized in Table 6.2. Saini, Shetty, and Giridhar (2014) reported the carotenoid content of fruits, flowers, and leaves of *M. oleifera* by liquid chromatography–mass spectrometry. They found that *M. oleifera* leaves have a higher carotenoid content (44.3–80.48 mg/100 g) followed by fruits and flowers (5.445–29.668 µg/g). Similarly, Saini, Shetty, et al. (2014) studied the carotenoid composition of *M. oleifera* leaves by HPLC; they reported that lyophilized leaves have higher total carotenoid content (208.49 mg/100 g) than fresh leaves (68.81 mg/100 g). On the other hand, Andarwulan et al. (2012) reported the carotenoid content of two *Moringa* species (*M. pterygosperma* and *M. citrifolia*), and found that *M. pterygosperma* has a total carotenoid content of 13.96 ± 0.19 mg/100 g and *M. citrifolia* has 3.28 ± 0.21 mg/100 g.

6.4 OTHER PHYTOCHEMICALS FROM *MORINGA* SPECIES

The diversity of the different types of phytochemical agents present in different *Moringa* species is very wide. In addition to polyphenolic and carotenoid compounds, other compounds such as glycosides, alkaloids, and tocopherols, among others, may be present. In different studies carried out, it has been found that the different

TABLE 6.2
Carotenoids from *Moringa* species

Species	Plant part	Compounds	Identification technique	Reference
Moringa citrifolia	Leaves	β-Carotene	HPLC	Andarwulan et al. (2012)
Moringa oleifera	Flowers Fruits Leaves	All-*E*-β-carotene, all-*E*-luteoxanthin, 13-Z-lutein, all-*E*-lutein, all-*E*-zeaxanthin and 15-Z-β-carotene	LC-MS	Saini, Shetty, and Giridhar (2014)
	Leaves	Lutein, β-carotene, *trans*-luteoxanthin, 13-*cis*-lutein, *trans*-lutein, *trans*-zeaxanthin, 15-*cis*-β-carotene, and *trans*-β-carotene	HPLC	Saini, Harish Prashanth, et al. (2014); Saini, Shetty, et al. (2014)
Moringa pterygosperma	Leaves	β-Carotene	HPLC	Andarwulan et al. (2012)

LC = liquid chromatography; MS = mass spectrometry; HPLC = high-performance liquid chromatography.

Moringa species contain higher concentrations of vitamins A and C than carrots and oranges, respectively (Singh et al. 2020). The amount of these metabolites present in *Moringa* species will vary by the environment in which these species grow and the type of extraction used. In the specific case of *Moringa oleifera*, different types of glucosinolates, alkaloids, and phytosterols have been found (Fernandes et al. 2020; Singh et al. 2020; Sodvadiya et al. 2020; Saini et al. 2016). However, phytochemicals such as alkaloids, terpenoids, and glucosinolates have been found in *M. concanensis*, *M. oleifera*, *M. ovalifolia*, and *M. peregrina*. Table 6.3 shows the different phytochemicals that were found in the different plant parts as well as different *Moringa* species.

6.5 PERSPECTIVES

There are currently around 13 species of *Moringa* distributed around the world. Although the most studied species is *M. oleifera*, some other species have been considered potential sources of micronutrients and phytochemicals (Özcan 2020; Padayachee and Baijnath 2020). *Moringa* flowers, leaves, and seeds are important sources of alkaloids, phenolic compounds, and tocopherols. In this sense, phytochemical *Moringa* compounds have antioxidant, anti-inflammatory, hypotensive, chemopreventive, and antibacterial properties (Ibrahim and Ghareeb 2020; Özcan 2020; Prabu et al. 2019; Saleem et al. 2020; Xu et al. 2019); also, it has been reported that these compounds provide glucose tolerance and decrease cholesterol absorption (Chhikara et al. 2020; Prabu et al. 2019; Toma et al. 2014; Ullah et al. 2015).

TABLE 6.3
Other phytochemicals in *Moringa* species

Species	Plant part	Compounds	Identification technique	Reference
Moringa concanensis	Flowers Leaves Seeds	Alkaloids, terpenoids	Phytochemical screening	Santhi and Sengottuvel (2016)
Moringa oleifera	Flowers	Glycoside (i.e., pterygospemin), β-sitosterol	HPLC	Chhikara et al. (2020); Fernandes et al. (2020); Kalappurayil and Joseph (2016)
	Leaves	Glucosinolates (i.e., 4-α-L-ramnosyloxy-benzyl)-isothiocyanate), acid caffeoylquinic, glycoside (i.e., niazimicin), isothiocyanate	Phytochemical screening and LC-MS	Godinez-Oviedo et al. (2016); Lim et al. (2020); Mohamad Shariff et al. (2020); Muhammad et al. (2016); Singh et al. (2020); Sodvadiya et al. (2020)
	Pods	Niazimixin, glucomorimgim, β-sitosterol, isothiocyanate	NMR and TLC	Chhikara et al. (2020); Padayachee and Baijnath (2020)
	Seeds	Phytosterol (β-sitosterol), glycoside (i.e., glycerol-1-(9-octadecaneate), lectins, isothiocyanate, alkaloids (glucomoringin)	NMR, ESI-MS, HPLC	Abbassy et al. (2018); Atolani et al. (2020); Chhikara et al. (2020); de Oliveira et al. (2020); Fernandes et al. (2020); Singh et al. (2020)
Moringa ovalifolia	Flowers Leaves Seeds	Alkaloids, terpenoids	Phytochemical screening	Preez et al. (2017) Santhi and Sengottuvel (2016)
Moringa peregrina	Leaves	Alkaloids, steroids (lupeol), glucosinolates (i.e., 4-α-L-ramnosyloxy-benzyl), isothiocyanate, phytosterols, thymol, ascorbic acid, retinol, phytol, xylene	Phytochemical screening, HPLC, GC-MS	Abd Rani et al. (2018); Ahl et al. (2017); Asghari et al. (2015); Bawadekji et al. (2019); Mansour et al. (2019); Senthilkumar et al. (2018;);Tahir et al. (2020); Ullah et al. (2015)
	Seeds	Isothiocyanate, β-sitosterol, alkaloids, glycoside	Phytochemical screening	Al Ameri et al. (2014); Saleem et al. (2020); Tahir et al. (2020)

| *Moringa stenopetala* | Leaves | Alkaloids, steroids, glycosides, phytosterols, isothiocyanate | Phytochemical screening | Fekadu et al. (2017); Geleta, Makonnen, Debella, Abebe, et al. (2016); Geleta, Makonnen, Debella, and Tadele (2016); Metsopkeng et al. (2020); Tamrat et al. (2017) |
| | Seeds | 4-(α-L-Rhamnopyranosyloxy) benzyl glucosinolate (glucomoringin), isothiocyanate, alkaloids (5,5-dimethyloxazolidine-2-thione) | TLC, NMR | Abd Rani et al. (2018); Adane et al. (2019) |

HPLC = high-performance liquid chromatography; LC = liquid chromatography; MS = mass spectrometry; NMR = nuclear magnetic resonance; TLC = thin layer chromatography; ESI-MS = electrospray ionization mass spectrometry; GC = gas chromatography.

Even though more *in vivo* studies are lacking to demonstrate the potential of the different compounds isolated from *Moringa* species (Saleem et al. 2020; Singh et al. 2020; Sodvadiya et al. 2020; Tahir et al. 2020), studies of all parts of the plant have gained greater interest as a potential source of phytochemicals with health promotion effects (Chhikara et al. 2020; Guzmán-Maldonado et al. 2020; Özcan 2020).

6.6 CONCLUSION

There are different *Moringa* species worldwide, commonly used to treat different ailments in folk medicine. Recent studies have shown that *Moringa* species are sources of phytochemical compounds such as phenolics, flavonoids, carotenoids, and alkaloids. Also, the different parts of the *Moringa* plant, such as bark, leaves, roots, and seeds, have a different profile of bioactive compounds and could be a promising source of novel therapeutic drugs.

REFERENCES

Abbassy, M.M.S., M.Z.M. Salem, N.M. Rashad, S.M. Afify, and A.Z.M. Salem. (2018). Nutritive and biocidal properties of agroforestry trees of *Moringa oleifera* Lam., *Cassia fistula* L., and *Ceratonia siliqua* L. as non-conventional edible vegetable oils. *Agroforestry Systems* 94(4): 1567–1579. https://doi.org/10.1007/s10457-018-0325-4.

Abd Rani, N.Z., K. Husain, and E. Kumolosasi. (2018). *Moringa* Genus: a review of phytochemistry and pharmacology. *Frontiers in Pharmacology* 9: 108. https://doi.org/10.3389/fphar.2018.00108.

Adane, L., M. Teshome, and Y. Tariku. (2019). Isolation of compounds from root bark extracts of *Moringa stenopetala* and evaluation of their antibacterial activities. *Journal of Pharmacognosy and Phytochemistry* 8(3): 4228–4244.

Ahl, H.A.H.S., W.M. Hikal, and A.A. Mahmound. (2017). Biological activity of *Moringa peregrina*: a review. *American Journal of Food Science and Health* 3(4); 83–87.

Al-Owaisi, M., N. Al-Hadiwi, and S.A. Khan. (2014). GC-MS analysis, determination of total phenolics, flavonoid content and free radical scavenging activities of various crude extracts of *Moringa peregrina* (Forssk.) Fiori leaves. *Asian Pacific Journal of Tropical Biomedicine* 4(12): 964–970. https://doi.org/10.12980/APJTB.4.201414B295.

Al Ameri, S.A., F.Y. Al Shaibani, A.J. Cheruth, A.I. Al-Awad, M.A.S. Al-Yafei, K. Karthishwaran, et al. (2014). Comparative phytochemical analysis of *Moringa oleifera* and *Moringa peregrina*. *Pharmacologyonline* 3: 216–221.

Al Juhaimi, F., K. Ghafoor, I.A.M. Ahmed, E.E. Babiker, and M.M. Özcan. (2017). Comparative study of mineral and oxidative status of *Sonchus oleraceus*, *Moringa oleifera* and *Moringa peregrina* leaves. *Journal of Food Measurement and Characterization* 11(4): 1745–1751. https://doi.org/10.1007/s11694-017-9555-9.

Andarwulan, N., D. Kurniasih, R.A. Apriady, H. Rahmat, A.V. Roto, and B.W. Bolling. (2012). Polyphenols, carotenoids, and ascorbic acid in underutilized medicinal vegetables. *Journal of Functional Foods* 4(1): 339–347. https://doi.org/10.1016/j.jff.2012.01.003.

Anunciacao, K.D., L.R.D. Sousa, T.R. Amparo, G.H.B. de Souza, P.M.D. Vieira, V.M.R. dos Santos, et al. (2020). Antioxidant activity evaluation and total phenolics in oils of seed extracted of the *Moringa oleifera* Lam. *Revista Virtual De Quimica* 12(1): 148–154. https://doi.org/10.21577/1984-6835.20200012.

Asghari, G., A. Palizban, and B. Bakhshaei. (2015). Quantitative analysis of the nutritional components in leaves and seeds of the Persian *Moringa peregrina* (Forssk.) Fiori. *Pharmacognosy Research* 7(3): 242–8. https://doi.org/10.4103/0974-8490.157968.

Atolani, A., O.E. Olorundare, P. Banerjee, O. Osin, R. Preissner, and A.A. Njan (2020). Isolation, characterisation and in silico toxicity evaluations of thiocarbamates, isothiocyanates, nitrile, glucosinolate and lipids from *Moringa oleifera* Lam. seed. *Journal of the Turkish Chemical Society Section A: Chemistry* 7: 233–242. https://doi.org/10.18596/jotcsa.569960.

Bawadekji, A., M. Imran, M.A.U. Mridha, and M. Al Ali. (2019). Phytochemical and anti-microbial activity evaluation of the water immiscible solvent extracts of *Moringa*. *Journal of Pure and Applied Microbiology* 13(3): 1483–1488. https://doi.org/10.22207/jpam.13.3.19.

Bennour, N., H. Mighri, H. Eljani, T. Zammouri, and A. Akrout. (2020). Effect of solvent evaporation method on phenolic compounds and the antioxidant activity of *Moringa oleifera* cultivated in Southern Tunisia. *South African Journal of Botany* 129: 181–190. https://doi.org/10.1016/j.sajb.2019.05.005.

Chhikara, N., A. Kaur, S. Mann, M.K. Garg, S.A. Sofi, and A. Panghal. (2020). Bioactive compounds, associated health benefits and safety considerations of *Moringa oleifera* L.: an updated review. *Nutrition & Food Science*: 24. https://doi.org/10.1108/nfs-03-2020-0087. <Go to ISI>://WOS:000535931800001.

de Oliveira, A.P.S., A.C. Agra-Neto, E.V. Pontual, T. de Albuquerque Lima, K. Cardoso, V. Cruz, et al. (2020). Evaluation of the insecticidal activity of *Moringa oleifera* seed extract and lectin (WSMoL) against *Sitophilus zeamais*. *Journal of Stored Products Research* 87. https://doi.org/10.1016/j.jspr.2020.101615.

Dehshahri, S., M. Wink, S. Afsharypuor, G. Asghari, and A. Mohagheghzadeh. (2012). Antioxidant activity of methanolic leaf extract of *Moringa peregrina* (Forssk.) Fiori. *Research in Pharmaceutical Sciences* 7(2): 111–118.

Dhakad, A.K., M. Ikram, S. Sharma, S. Khan, V.V. Pandey, and A. Singh. (2019). Biological, nutritional, and therapeutic significance of *Moringa oleifera* Lam. *Phytotherapy Research* 33(11): 2870–2903. https://doi.org/10.1002/ptr.6475.

Elbatran, S.A., O.M. Abdel-Salam, K.A. Abdelshfeek, N.M. Nazif, S.I. Ismail, and F.M. Hammouda. (2005). Phytochemical and pharmacological investigations on *Moringa peregrina* (Forssk) Fiori. *Natural Product Sciences* 11(4): 199–206.

Fejer, J., I. Kron, V. Pellizzeri, M. Pluchtova, A. Eliasova, L. Campone, et al. (2019). First report on evaluation of basic nutritional and antioxidant properties of *Moringa oleifera* Lam. from Caribbean island of Saint Lucia. *Plants – Basel* 8(12): 15. https://doi.org/10.3390/plants8120537.

Fekadu, N., H. Basha, A. Meresa, S. Degu, B. Girma, and B. Geleta. (2017). Diuretic activity of the aqueous crude extract and hot tea infusion of *Moringa stenopetala* (Baker f.) Cufod. leaves in rats. *Journal of Experimental Pharmacology* 9: 73–80.

Fernandes, A., A. Bancessi, J. Pinela, M.I. Dias, A. Liberal, R.C. Calhelha, et al. (2020). Nutritional and phytochemical profiles and biological activities of *Moringa oleifera* Lam. edible parts from Guinea-Bissau (West Africa). *Food Chemistry* 341(Pt 1): 128229. https://doi.org/10.1016/j.foodchem.2020.128229.

Geleta, B., E. Makonnen, A. Debella, A. Abebe, and N. Fekadu. (2016). *In vitro* vasodilatory activity and possible mechanisms of the crude extracts and fractions of *Moringa stenopetala* (Baker f.) Cufod. leaves in isolated thoracic aorta of guinea pigs. *Journal of Experimental Pharmacology* 8: 35–42. https://doi.org/10.2147/JEP.S117545.

Geleta, B., E. Makonnen, A. Debella, and A. Tadele. (2016). *In vivo* antihypertensive and antihyperlipidemic effects of the crude extracts and fractions of *Moringa stenopetala*

(Baker f.) Cufod. leaves in rats. *Frontiers in Pharmacology* 7: 97. https://doi.org/10.3389/fphar.2016.00097.

Gupta, S., R. Jain, S. Kachhwaha, and S.L. Kothari. (2018). Nutritional and medicinal applications of *Moringa oleifera* Lam. – review of current status and future possibilities. *Journal of Herbal Medicine* 11: 1–11. https://doi.org/10.1016/j.hermed.2017.07.003.

Godinez-Oviedo, A., N. Guemes-Vera, and O.A Acevedo-Sandoval. (2016). Nutritional and phytochemical composition of *Moringa oleifera* Lam and its potential use as nutraceutical plant: a review. *Pakistan Journal of Nutrition* 15(4): 397–405.

Guzmán-Maldonado, S.H., M.J. López-Manzano, T.J. Madera-Santana, C.A. Núñez-Colín, C.P. Grijalva-Verdugo, A.G. Villa-Lerma, et al. (2020). Nutritional characterization of *Moringa oleifera* leaves, seeds, husks and flowers from two regions of Mexico. *Agronomia Colombiana* 38(2): 189–199. https://doi.org/10.15446/agron.colomb.v38n2.82644.

Ibrahim, A.M., and M.A. Ghareeb. (2020). Preliminary phytochemical screening, total phenolic content, *in vitro* antioxidant and molluscicidal activities of the methanolic extract of five medicinal plants on *Biomphalaria alexandrina* snails. *Journal of Herbs, Spices and Medicinal Plants* 26(1): 40–48. https://doi.org/10.1080/10496475.2019.1666769.

Kalappurayil, T.M., and B.P. Joseph. (2016). A review of pharmacognostical studies on *Moringa oleifera* Lam. flowers. *Pharmacognosy Journal* 9(1): 1–7. https://doi.org/10.5530/pj.2017.1.1.

Lim, W.F., M.I. Mohamad Yusof, L.K. Teh, and M.Z. Salleh. (2020). Significant decreased expressions of CaN, VEGF, SLC39A6 and SFRP1 in MDA-MB-231 xenograft breast tumor mice treated with *Moringa oleifera* leaves and seed residue (MOLSr) extracts. *Nutrients* 12(10). https://doi.org/10.3390/nu12102993.

Makita, C., L. Chimuka, E. Cukrowska, P.A. Steenkamp, M. Kandawa-Schutz, A.R. Ndhlala, et al. (2017). UPLC-qTOF-MS profiling of pharmacologically important chlorogenic acids and associated glycosides in *Moringa ovalifolia* leaf extracts. *South African Journal of Botany* 108: 193–199. https://doi.org/10.1016/j.sajb.2016.10.016.

Makita, C., L. Chimuka, P. Steenkamp, E. Cukrowska, and E. Madala. (2016). Comparative analyses of flavonoid content in *Moringa oleifera* and *Moringa ovalifolia* with the aid of UHPLC-qTOF-MS fingerprinting. *South African Journal of Botany* 105: 116–122. https://doi.org/10.1016/j.sajb.2015.12.007.

Mansour, M., M.F. Mohamed, A. Elhalwagi, H.A. El-Itriby, H.H. Shawki, and I.A. Abdelhamid. (2019). *Moringa peregrina* leaves extracts induce apoptosis and cell cycle arrest of hepatocellular carcinoma. *Biomedical Research International* 2019: 2698570. https://doi.org/10.1155/2019/2698570.

Metsopkeng, C.S., M.E. Nougang, P.A. Nana, A.T. Arfao, P.N. Bahebeck, C.L. Djimeli, et al. (2020). Comparative study of *Moringa stenopetala* root and leaf extracts against the bacteria *Staphyloccocus aureus* strain from aquatic environment. *Scientific African* 10. https://doi.org/10.1016/j.sciaf.2020.e00549.

Mohamad Shariff, N.F. Syazleen, T. Singgampalam, C.H. Ng, and C.S. Kue. (2020). Antioxidant activity and zebrafish teratogenicity of hydroalcoholic *Moringa oleifera* L. leaf extracts. *British Food Journal* 122(10): 3129–3137. https://doi.org/10.1108/bfj-02-2020-0113.

Moyo, B., S. Oyedemi, P.J. Masika, and V. Muchenje. (2012). Polyphenolic content and antioxidant properties of *Moringa oleifera* leaf extracts and enzymatic activity of liver from goats supplemented with *Moringa oleifera* leaves/sunflower seed cake. *Meat Science* 91(4): 441–447. https://doi.org/10.1016/j.meatsci.2012.02.029.

Muhammad, H.I., M.Z. Asmawi, and N.A.K. Khan. (2016). A review on promising phytochemical, nutritional and glycemic control studies on *Moringa oleifera* Lam. in tropical and sub-tropical regions. *Asian Pacific Journal of Tropical Biomedicine* 6(10): 896–902. https://doi.org/10.1016/j.apjtb.2016.08.006.

Oldoni, T.L.C., N. Merlin, M. Karling, S.T. Carpes, S.M. de Alencar, R.G.F. Morales, et al. (2019). Bioguided extraction of phenolic compounds and UHPLC-ESI-Q-TOF-MS/MS characterization of extracts of *Moringa oleifera* leaves collected in Brazil. *Food Research International* 125: 9. https://doi.org/10.1016/j.foodres.2019.108647.

Özcan, M.M. (2020). *Moringa* spp: composition and bioactive properties. *South African Journal of Botany* 129: 25–31. https://doi.org/10.1016/j.sajb.2018.11.017.

Padayachee, B., and H. Baijnath. (2020). An updated comprehensive review of the medicinal, phytochemical and pharmacological properties of *Moringa oleifera*. *South African Journal of Botany* 129: 304–316. https://doi.org/10.1016/j.sajb.2019.08.021.

Prabu, S.L., A. Umamaheswari, and A. Puratchikody. (2019). Phytopharmacological potential of the natural gift *Moringa oleifera* Lam and its therapeutic application: an overview. *Asian Pacific Journal of Tropical Medicine* 12(11): 485–498. https://doi.org/10.4103/1995-7645.271288.

Preez, I. du, R.-J. Bussel, and D. Mumbengegw. (2017). Evaluation of the antiplasmodial properties of namibian medicinal plant species, *Moringa ovalifolia*. *Research Journal of Medicinal Plants* 11(4): 167–173. https://doi.org/10.3923/rjmp.2017.167.173.

Saini, R.K., K.V.H. Prashanth, N.P. Shetty, and P. Giridhar. (2014). Elicitors, SA and MJ enhance carotenoids and tocopherol biosynthesis and expression of antioxidant related genes in *Moringa oleifera* Lam. leaves. *Acta Physiologiae Plantarum* 36(10): 2695–2704. https://doi.org/10.1007/s11738-014-1640-7.

Saini, R.K., N.P. Shetty, and P. Giridhar. (2014). Carotenoid content in vegetative and reproductive parts of commercially grown *Moringa oleifera* Lam. cultivars from India by LC-APCI-MS. *European Food Research and Technology* 238(6): 971–978. https://doi.org/10.1007/s00217-014-2174-3.

Saini, R.K., N.P. Shetty, M. Prakash, and P. Giridhar. (2014). Effect of dehydration methods on retention of carotenoids, tocopherols, ascorbic acid and antioxidant activity in *Moringa oleifera* leaves and preparation of a RTE product. *Journal of Food Science and Technology* 51(9): 2176–2182. https://doi.org/10.1007/s13197-014-1264-3.

Saini, R.K., I. Sivanesan, and Y.S. Keum. (2016). Phytochemicals of *Moringa oleifera*: a review of their nutritional, therapeutic and industrial significance. *3 Biotech* 6(2): 203. https://doi.org/10.1007/s13205-016-0526-3. www.ncbi.nlm.nih.gov/pubmed/28330275.

Saleem, A., M. Saleem, and M.F. Akhtar. (2020). Antioxidant, anti-inflammatory and antiarthritic potential of *Moringa oleifera* Lam: an ethnomedicinal plant of Moringaceae family. *South African Journal of Botany* 128: 246–256. https://doi.org/10.1016/j.sajb.2019.11.023.

Santhi, K., and R. Sengottuvel. (2016). Qualitative and quantitative phytochemical analysis of *Moringa concanensis* Nimmo. *International Journal of Current Microbiology and Applied Sciences* 5(1): 633–640. https://doi.org/10.20546/ijcmas.2016.501.064.

Senthilkumar, A., N. Karuvantevida, L. Rastrelli, S.S. Kurup, and A.J. Cheruth. (2018). Traditional uses, pharmacological efficacy, and phytochemistry of *Moringa peregrina* (Forssk.) Fiori. – a review. *Frontiers in Pharmacology* 9: 465. https://doi.org/10.3389/fphar.2018.00465.

Singh, A.K., H.K. Rana, T. Tshabalala, R. Kumar, A. Gupta, A.R. Ndhlala, et al. (2020). Phytochemical, nutraceutical and pharmacological attributes of a functional crop *Moringa oleifera* Lam: an overview. *South African Journal of Botany* 129: 209–220. https://doi.org/10.1016/j.sajb.2019.06.017.

Sodvadiya, M., H. Patel, A. Mishra, and S. Nair. (2020). Emerging insights into anticancer chemopreventive activities of nutraceutical *Moringa oleifera*: molecular mechanisms, signal transduction and *in vivo* efficacy. *Current Pharmacology Reports* 6(2): 38–51. https://doi.org/10.1007/s40495-020-00210-z.

Tahir, N.A., H.O. Majeed, H.A. Azeez, J.M. Faraj, and W.R.M. Palani. (2020). Allelopathic plants: 27. *Moringa* species. *Allelopathy Journal* 50(1): 35–48. https://doi.org/10.26651/allelo.j/2020-50-1-1272.

Tamrat, Y., T. Nedi, S. Assefa, T. Teklehaymanot, and W. Shibeshi. (2017). Anti-inflammatory and analgesic activities of solvent fractions of the leaves of *Moringa stenopetala* Bak. (Moringaceae) in mice models. *BMC Complementary and Alternative Medicine* 17(1): 473. https://doi.org/10.1186/s12906-017-1982-y.

Toma, A., E. Makonnen, Y. Mekonnen, A. Debella, and S. Addisakwattana. (2014). Intestinal α-glucosidase and some pancreatic enzymes inhibitory effect of hydroalcoholic extract of *Moringa stenopetala* leaves. *BMC Complementary and Alternative Medicine* 14. https://doi.org/10.1186/1472-6882-14-180.

Ullah, M.F., S.H. Bhat, and F.M. Abuduhier. (2015). Antidiabetic potential of hydro-alcoholic extract of *Moringa peregrine* leaves: implication as functional food for prophylactic intervention in prediabetic stage. *Journal of Food Biochemistry* 39(4): 360–367. https://doi.org/10.1111/jfbc.12140.

Vergara-Jimenez, M., M.M. Almatrafi, and M.L. Fernandez. (2017). Bioactive components in *Moringa oleifera* leaves protect against chronic disease. *Antioxidants* 6(4). https://doi.org/10.3390/antiox6040091.

Xu, Y.B., G.L. Chen, and M.Q. Guo. (2019). Antioxidant and anti-inflammatory activities of the crude extracts of *Moringa oleifera* from Kenya and their correlations with flavonoids. *Antioxidants* 8(8): 12. https://doi.org/10.3390/antiox8080296.

Zhao, B.B., J.W. Deng, H. Li, Y.Q. He, T. Lan, D. Wu, et al. (2019). Optimization of phenolic compound extraction from Chinese *Moringa oleifera* leaves and antioxidant activities. *Journal of Food Quality* 2019: 13. https://doi.org/10.1155/2019/5346279.

7 Peptides
The Other Bioactive Constituents of Moringa

Sara Avilés-Gaxiola
Centro de Investigación en Alimentación y Desarrollo A.C.,
Carretera a Eldorado Km. 5.5, Col. Campo El Diez, Culiacán,
Sinaloa, 80110 México.

J. Basilio Heredia
Centro de Investigación en Alimentación y Desarrollo A.C.,
Carretera a Eldorado Km. 5.5, Col. Campo El Diez, Culiacán,
Sinaloa, 80110 México.
Corresponding author: jbheredia@ciad.mx

CONTENTS

7.1 *MORINGA OLEIFERA* PROTEIN GENERALITIES

Moringa oleifera (MO) is an angiosperm plant that belongs to the Moringaceae family. This family is made up of only 13 species. Among these species, *M. oleifera*, commonly known as 'the tree of life', 'miracle tree', and 'mother's best friend', has been the most studied. All parts of MO are suitable for human and animal intake, and the consumption and usage of these plant parts, especially the leaves and seeds, have been increasing due to their high protein content (Su and Chen 2020). The protein content of the different parts of the MO tree is detailed in Table 7.1.

The protein profile of MO leaves is mainly composed of proteins over 29 kDa and others in the range of 14–20 kDa (Paula et al. 2017). MO leaves have been proposed as

TABLE 7.1
Protein content of *Moringa oleifera* tree parts

Tree part	Protein content (g/100 g)	Reference
Leaf	19.3–35.0[1]	Olson et al. (2016)
Seed	18.9–37.2	Ijarotimi et al. (2013); Bridgemohan et al. (2014)
Pod	18.4–19.8[1]	(Abdulkadir et al. (2016); Gidamis et al. (2003)
Root	14.0–16.9	Braide et al. (2017)
Flower	17.8–25.2[1]	Arise et al. (2014); Madane et al. (2019)

[1] Dry weight.

a protein supplement, given their amino acid composition, which contains the 10 essential amino acids: tyrosine, methionine, threonine, phenylalanine, valine, leucine, isoleucine, histidine, tryptophan, and lysine. As for lysine content, MO leaves have eight times than cornmeal and more methionine than other leaves like alfalfa, although it is deficient in cystine (Aderinola et al. 2018; Su and Chen 2020). Due to its high protein content, MO leaf powder is considered an alternative supplement to reduce anemia and, therefore, malnutrition in low-income children. It has been found that supplementation with MO leaves reduced the prevalence of moderate and severe anemia by 68.2% and 77.9%, respectively. So, its use as part of the formulation of infant formulas and food products such as scholar snacks, juices, and yogurt would have a great positive effect (Dewey et al. 2009; Glover-Amengor et al. 2017; Hekmat et al. 2015; Saa et al. 2019).

The MO seed protein profile is composed of proteins of 29, 14.2, and 6.5 kDa, which are mainly coagulating proteins (Jain et al. 2019). MO seed protein has a high content of essential amino acids, which indicates its high nutritional value. It has high contents of valine and hydrophobic amino acids like leucine and alanine, while tryptophan is at a low concentration (Aderinola et al. 2018). Hydrophobic amino acids contribute to the antioxidant activity of MO seed protein because they help to improve their solubility in lipid medium, like cellular membranes (Rajapakse et al. 2005). MO seed flour has been included in infant porridge formulation (Saa et al. 2019). However, due to the bitterness of MO seed flour, its utilization is still limited. In terms of digestibility, *in vitro* studies have revealed that MO leaf proteins are more easily digested and absorbed than seed proteins (Mune Mune et al. 2016). As for the other parts of the MO tree, studies on identifying its amino acid profile are still lacking. In Table 7.2, the amino acid composition of MO leaves and seed protein is detailed.

7.2 *MORINGA OLEIFERA* PEPTIDE GENERATION, ISOLATION, AND IDENTIFICATION

Peptides are protein fragments ranging from 2 to 50 amino acids linked by peptide bonds. Some are naturally found in plant sources, and they can also be produced by protein fermentation or enzymatic hydrolysis (Apone et al. 2019). Peptides may act as antithrombotic, antihypertensive, immunomodulating, anticancer, and antioxidant agents. Plant peptides have been typically obtained from soy and pulses like chickpea, beans, lentils, and cereal sources, especially oat (Chakrabarti et al. 2018). The usage

TABLE 7.2
Amino acid composition of *Moringa oleifera* seed and leaves protein. The data correspond to percentages

Amino acid	Seed protein[1]	Leaves protein[2]
ASK	6.27	–
THR	3.35	4.44
SER	3.19	4.75
GLX	25.70	–
PRO	5.90	4.76
GLY	5.98	6.15
ALA	4.41	6.84
CYS	3.42	0.22
VAL	4.46	5.28
MET	1.93	2.69
ILE	3.00	4.40
LEU	6.33	8.54
TYR	2.27	4.08
PHE	5.35	8.14
HIS	2.95	4.71
LYS	1.29	5.52
ARG	13.19	6.20
TRP	0.99	1.22
ASP	–	7.83
GLU	–	14.22

[1] Aderinola et al. (2018), [2] Moyo et al. (2011).

of legumes and cereal proteins is somewhat limited. As for legumes, it has been found that their derived peptides may maintain the allergenic fractions of various proteins (Belsito et al. 2017). On the other hand, the market acceptance of cereal proteins is low because trace levels of gluten can be found even in gluten-free cereals such as rice, maize, quinoa, buckwheat, millet, or sorghum. This contamination may occur at different steps during their cultivation and processing, transportation, or handling. Recent research has focused on obtaining peptides from other plant sources, such as leafy plants (Bustamante et al. 2017).

Two naturally occurring peptides have been isolated from MO leaves. They were designated under the name of morintides 1 and 2, both 44 amino acids long. Both shared 50% of their structure, composed of eight residues of cysteine, six residues of glycine, five residues of asparagine, and five glutamine residues. Morintide 1 contains four disulfide bonds within its structure, being very stable (Kini et al. 2017). The naturally occurring peptide, trypsin inhibitor, has also been isolated from MO flower (Pontual et al. 2009, 2018; Shebek et al. 2015). Plants produce trypsin inhibitors to protect themselves from pest attacks, but also, different plant trypsin inhibitors are bioactive (Avilés-Gaxiola et al. 2018).

MO peptides have been produced from seeds, leaves flower total protein and seed globulin (Aderinola et al. 2019; Garza et al. 2017; Shakir et al. 2019). MO seed protein has been obtained by the alkaline extraction and isoelectric precipitation technique, and it has been found that the pH with the highest yield is 11 for alkaline extraction and 4 for isoelectric precipitation (Garza et al. 2017). Although this method is very efficient for legume protein extraction, the yield was low for MO seed, mainly due to the native content and biochemical characteristics of its protein (Garza et al. 2017). MO seed globulin has been obtained through dialysis (Aderinola et al. 2019).

On the other hand, MO leaves and flower proteins have been obtained by ammonium sulfate or acetone precipitation followed by dialysis and alkali extraction and acid precipitation with a purity of 81.12% (Dahot 1998; Huang et al. 2020; Shakir et al. 2019). MO peptides have been produced by enzymatic hydrolysis. Enzymatic hydrolysis is the most common and effective way of producing bioactive peptides (Rayaprolu et al. 2017). In this process, a combination of enzymes, or an enzyme alone, is used. Each enzyme has a specific cutting site within the protein. This factor, along with particular process parameters such as pH, temperature, time, and enzyme(s) concentration, influence the degree of hydrolysis (Zambrowicz et al. 2013). The enzymes that have been used for the generation of MO peptides are trypsin, chymotrypsin, pepsin, alcalase, bromelain, flavourzyme, dispase, papain, and protamex (Aderinola et al. 2019; Garza et al. 2017; Huang et al. 2020; Liang et al. 2019; Yun et al. 2020). Table 7.3 shows the enzymes used for the production of bioactive peptides from MO.

Once peptides are produced, further steps are needed to fractionate and/or identify them. First, digestive enzymes are heat-inactivated and subsequently removed by centrifugation. Ultrafiltration is the most common way to fractionate peptides. Few MO peptides have been isolated and identified. For this purpose, ultra-high performance liquid chromatography with quadrupole time-of-flight mass spectrometry (UPLC-Q-TOF-MS/MS) has been mostly used (Lin et al. 2019; Zhuang et al. 2016). Figure 7.1 shows a general diagram with the recommended steps to obtain and characterize *MO* bioactive peptides.

7.3 BIOLOGICAL POTENTIAL OF *MORINGA OLEIFERA* PEPTIDES

7.3.1 Antioxidant Capacity

The MO plant has been used for its culinary properties and its medicinal effects, especially its antioxidant potential. MO phytochemicals, including peptides, are promising molecules to counteract the undesirable effect of oxidative stress. Oxidative stress is a major factor in developing and progressing diseases like cancer and cardiovascular and neurodegenerative diseases (Loizzo and Tundis 2019).

Ten antioxidant peptides were isolated from MO seed protein hydrolyzed by flavor protease. Their sequences were identified as follows: GY, PFE, YTR, FG, QY, IN, SF, SP, YFE, and IY. They exhibited strong scavenging activities on DPPH with EC_{50} of 2.28, 1.60, 1.77, 2.15, 0.97, 1.30, 0.75, 0.91, 1.21 and 0.79 mg/l, respectively. Also, they exhibited antioxidant activity on ABTS with EC_{50} of 1.03, 0.84, 0.95, 0.65, 0.37, 0.54, 0.33, 0.36, 0.67 and 0.32 and mg/ml, respectively. The peptides SF and QY protected liver cells from oxidative damage induced by the radical H_2O_2 by increasing superoxide dismutase activity by around 66% and 88%, respectively. Both peptides increased catalase activity twofold (Liang et al. 2020). On the other hand,

TABLE 7.3
Enzymes used for the production of bioactive peptides from *Moringa oleifera*

Enzyme	Origin	Optimal pH	Optimal temperature (°C)	Cutting site	Reference
Alcalase	*Bacillus licheniformis*	6.5–8.5	60	Hydrolyzes most of the peptide bonds within a protein molecule	Daud et all (2013)
Chymotrypsin	Bovine pancreas	7.8	25	Cleaves to the C-terminal side of tryptophan, tyrosine, phenylalanine, and methionine	Szabó et al. (2015)
Papain	Papaya	5.0–7.0	65	Cleaves to peptide bonds of leucine and glycine	Mótyán et al. (2013)
Pepsin	Pig gastric mucosa	2.0	37.0	Cleaves after phenylalanine and leucine	Ahn et al. (2013)
Trypsin	Porcine pancreas	7.6	25	Cleaves to C-terminal of lysine and arginine	Walmsley et al. (2013)
Bromelain	Pineapple stem	4.6	25	Cleaves to C-terminal of lysine, alanine, tyrosine, and glycine	Arshad et al. (2014)
flavourzyme	*Aspergillus oryzae*	8.0	50	Hydrolyzes most of the peptide bonds within a protein molecule	Bruno et al. (2019)
Dispase	*Bacillus polymyxa*	7.5	37	Cleaves to leucine-phenylalanine and serine-phenylalanine bonds	Zhu et al. (2012)
Protamex	*Bacillus* sp.	8.0	60	–	Choi et al. (2009)

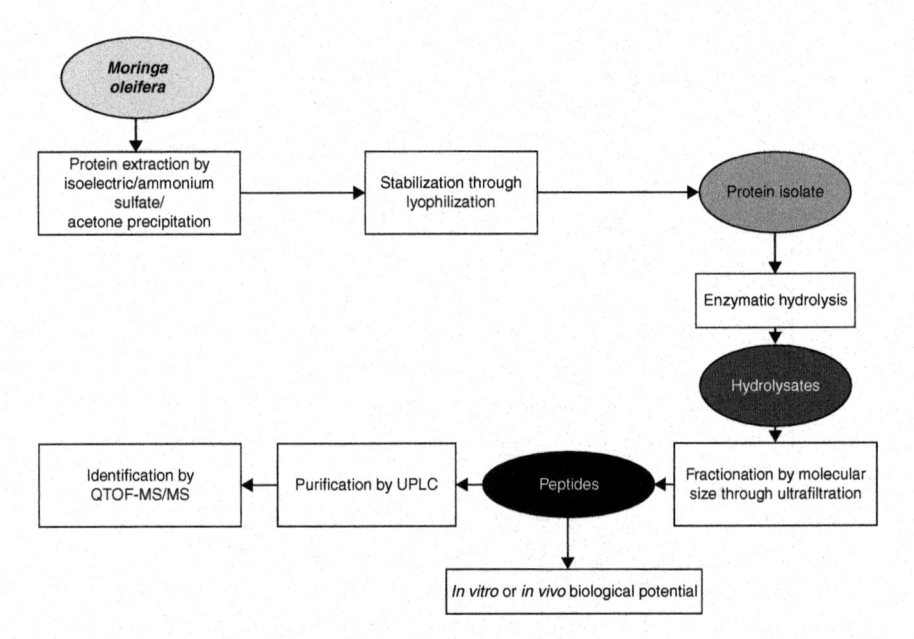

FIGURE 7.1 General diagram with the steps to obtain and characterize MO bioactive peptides.

MO seed hydrolysate produced with pepsin had high DPPH and ferric reducing antioxidant power (FRAP) activity at 36% and 0.04%, respectively, with 1 mg/ml concentration. This antioxidant activity is higher when compared with the protein isolate (Aderinola et al. 2018). Moreover, MO seed protein hydrolyzed fractions of <3.5 kDa with Flavourzyme had a strong antioxidant capacity with EC_{50} values of DPPH, •OH, ABTS, and $O_2^{•-}$ free radical scavenging rates of 4.0, 4.2, 5.3, and 4.3 mg/mLl respectively (Liang et al. 2019). MO seed hydrolysates using pepsin or trypsin had high scavenging activity and high ferric reducing power (AO et al. 2018).

MO seed peptides have also been produced from the globulin fraction; <1 kDa trypsin hydrolysate had a high antioxidant effect with 29.13% and 180% for DPPH and FRAP, respectively, when using peptides in a concentration of 1 mg/ml (Aderinola et al. 2019). MO leaf protein hydrolysates also have antioxidant activity. MO leaf protein hydrolyzed with alcalase had 65.97% DPPH inhibition, 1887.46 nM total antioxidant capacity (TAC) and 1413.28 µM FRAP (Yun et al. 2020). MO leaf submitted to gastrointestinal hydrolysis promoted the generation of eight antioxidant peptides PR, LIK, LIH, LIP, LILI, GE, VLI, and PF. They protected oxidatively damaged erythrocytes by inhibiting malondialdehyde formation, decreasing the hemolysis ratio from 48.45% to 10% at 25 mg/ml concentration. This hydrolysate had DPPH and oxygen radical absorbance capacity (ORAC) values of 600.16 µM, and 1839.71 TE/g (Lin et al. 2019). A protein isolate of MO leaves at 500 mg/kg of body weight reduced oxidative stress in diabetic mice by decreasing malondialdehyde level by 51.63% and an increase in catalase activity by 56.8% (Paula et al. 2017). For their high antioxidant activity, MO hydrolysates must be considered for further inclusion in the formulation of nutrition products, healthcare products, functional foods, beauty, and skincare products.

7.3.2 ANTIDIABETIC CAPACITY

MO leaves have been used to treat diabetes in traditional medicine; 500 mg/kg of body weight of the protein isolate of MO leaves reduced blood glucose level in diabetic mice by 34.4%, 60.9%, and 66.4% after 1, 3, and 5 h of administration, respectively. The effect may be due to the antidiabetic activity of the bioactive peptides produced after the *in vivo* gastrointestinal digestion of the MO leaf protein (Paula et al. 2017).

7.3.3 ANTIHYPERTENSIVE CAPACITY

Peptides with antihypertensive activity have been found encrypted within MO seed protein. The >10 kDa fraction of the MO seed protein hydrolyzed with pepsin–trypsin during 5 h inhibited angiotensin-converting enzyme with an IC_{50} of 0.224 µg/µl (Garza et al. 2017). The 1–3 kDa fraction of MO seed globulin hydrolysate obtained using trypsin also inhibited the angiotensin-converting enzyme by 72.48% at a concentration of 0.5 mg/ml (Aderinola et al. 2019).

7.3.4 ANTIFUNGAL ACTIVITY

Hevein is a 43-amino acid peptide with a chitin-binding domain. Chitin is a constituent of fungal cell walls, and therefore hevein-like-peptides play an important role in plant defense (Wong et al. 2017). Two 8C-hevein-like peptides isolated from MO leaves, named morintide 1 and morintide 2, can bind to chitin strongly and inhibit the growth of phytopathogenic fungi *Alternaria alternata* and *A. brassiciola* with an IC_{50} in the range of 25.5–60.43 µg/ml after incubation for 24 h. There are two mechanisms involved in morintide antifungal activity: (1) they were able to bind to nascent chitin fibers in newly formed chitin chains of growing hyphae, disrupting cell wall morphogenesis and stopping fungal growth; and (2) they easily penetrate through the cell wall of *A. alternata* and *A. brassiciola*, promoting the leakage of cytoplasm (Kini et al. 2017). *Alternaria* spp are human allergens that cause reactions that lead to asthma and cause around 20% of agricultural spoilage (Pinto and Patriarca 2017).

7.3.5 ANTIBACTERIAL ACTIVITY

Antimicrobial peptides are a powerful alternative for treating infections of bacterial strains resistant to multiple common antibiotics. The protein extracts of flowers and leaves of MO have antimicrobial activity against the multidrug-resistant bacteria *Staphylococcus aureus*, showing a maximum inhibition zone with a minimum inhibitory concentration of 166 and 500 µg/ml, respectively (Shakir et al. 2019).

Escherichia coli viability is affected by a cationic protein isolated from the seeds of MO. This protein fuses with the microbial inner and outer membranes of this bacteria, promoting its death (Shebek et al. 2015). The growth of *E. coli*, *S. aureus*, and *Bacillus subtilis* was inhibited by small proteins from MO leaves with an inhibitory zone of 10 mm in a dose of 0.80 mg/ml (Dahot 1998).

A peptide isolated from MO seeds, called 'flo', reduced by around 50% the cell survival of *E. coli* after a 2-h incubation. The antibacterial activity of this peptide is mainly due to a hydrophobic loop and α-helices within its structure and also to its positive charge that binds to the negative membrane of the bacteria, producing the formation of micelles and releasing the bacterial content (Fisch et al. 2004).

The MO protein and peptides possess antimicrobial activity that may be used as a potential drug against multidrug-resistant bacteria.

7.3.6 ANTIPROTOZOAL ACTIVITY

The MO flower trypsin inhibitor caused lysis of *Trypanosoma cruzi* with $LC_{50/24\,h}$ of 41.20 µg/ml (Pontual et al. 2018). *Trypanosoma cruzi* is a life-threatening parasite that causes trypanosomiasis, also known as Chagas disease. There are three phases of the disease: acute, indeterminate, and chronic. While symptoms last approximately 2 months in acute infections, in chronic infections they can last for years. Antiprotozoal peptides are a natural option for treating this disease (Carabarin-Lima et al. 2013).

7.3.7 ANTITERMITIC ACTIVITY

The naturally occurring trypsin inhibitor purified from the MO flower promoted the mortality of the termite *Nasutitermes corniger* workers, not affecting soldier survival (Pontual et al. 2009). The mechanism of the termiticidal activity of the MO flower trypsin inhibitor is not yet known.

Table 7.4 summarizes the most relevant information regarding the bioactive potential of MO peptides.

7.4 CONCLUSIONS

The MO tree is widely distributed in the world, which makes it an accessible raw material. The nutritional qualities of its different parts have been widely studied and reported, which has led to the identification of various nutraceutical compounds, as well as their function in terms of health. Peptides are among the most recently studied MO nutraceutical molecules. Little has been evaluated regarding them compared to other molecules; however, what has been found so far and reported in this chapter gives rise to continued study on their effect on various diseases, suggesting their future use both in the food and pharmaceutical industries. However, more studies *in vivo* are still lacking.

TABLE 7.4

Bioactivity of *Moringa oleifera* peptides

Moringa oleifera part	Method of extraction	Final product	Bioactivity	Bioactivity assay	References
Seed	Hydrolysis by flavor protease	GY, PFE, YTR, FG, QY, IN, SF, SP, YFE, and IY	Antioxidant	DPPH EC_{50} of 2.28, 1.60, 1.77, 2.15, 0.97, 1.30, 0.75, 0.91, 1.21, and 0.79 mg/l, respectively. ABTS EC_{50} of 1.03, 0.84, 0.95, 0.65, 0.37, 0.54, 0.33, 0.36, 0.67, 0.32, and 0.38 mg/ml, respectively	Liang et al. (2020)
Seed	Hydrolysis by flavor protease	SF and QY	Antioxidant	Increased the activity of superoxide dismutase and catalase in liver cells by 66% and 88%, respectively	Liang et al. (2020)
Seed	Hydrolysis by pepsin	Hydrolysate	Antioxidant	1 mg/ml of the hydrolysate had DPPH and FRAP activities of 36% and 0.04%, respectively	Aderinola et al. (2018)
Seed	Hydrolysis by flavourzyme	<3.5 kDa fraction	Antioxidant	EC_{50} values of DPPH, •OH, ABTS, and O_2•$^-$ scavenging rates of 4.0, 4.2, 5.3, and 4.3 mg/ml, respectively	Liang et al. (2019)
Seed	Globulin hydrolysis by trypsin	<1 kDa fraction	Antioxidant	29.13% and 180% for DPPH and FRAP, respectively, at a concentration of 1 mg/ml	Aderinola et al. (2019)
Leaf	Hydrolysis by alcalase	Hydrolysate	Antioxidant	65.97% of DPPH inhibition, 1887. 46 nM TAC, and 1413.28 μM FRAP	Yun et al. (2020)
Leaf	Gastrointestinal *in vitro* hydrolysis	PR, LIK, LIH, LIP, LILI, GE, VLI, and PF	Antioxidant	At a concentration of 25 mg/ml decreased the hemolysis ratio from 48.45% to 10% of oxidatively damaged erythrocytes by inhibition of malondialdehyde formation; DPPH and ORAC values of 600.16 μM and 1839.71 TE/g, respectively	Lin et al. (2019)
Leaf	Gastrointestinal *in vivo* hydrolysis	Hydrolysate	Antioxidant	500 mg/kg reduced oxidative stress in diabetic mice by a decrease in malondialdehyde level by 51.63% and an increase in catalase activity by 56.8%	Paula et al. (2017)

(continued)

TABLE 7.4 (Continued)
Bioactivity of *Moringa oleifera* peptides

Moringa oleifera part	Method of extraction	Final product	Bioactivity	Bioactivity assay	References
Leaf	Gastrointestinal *in vivo* hydrolysis	Hydrolysate	Antidiabetic	500 mg/kg reduced blood glucose level in diabetic mice by 34.4%, 60.9%, and 66.4% after 1, 3, and 5 h of its administration, respectively	Paula et al. (2017)
Seed	Hydrolysis by pepsin–trypsin	>10 kDa fraction	Antihypertensive	Inhibited angiotensin-converting enzyme with an IC_{50} of 0.224 µg/µl	Garza et al. (2017)
Seed	Globulin hydrolyzed by trypsin	Hydrolysate	Antihypertensive	Inhibited angiotensin-converting enzyme by 72.48% at a concentration of 0.5 mg/ml	Aderinola et al. (2019)
Leaves	Naturally occurring peptides	Two 8C-hevein-like peptides	Antifungal	Inhibited the growth of *Alternaria alternata* and *A. brassiciola* with an IC_{50} in the range of 25.5–60.43 µg/ml after incubation for 24 h	Kini et al. (2017)
Flower	–	Protein extract	Antibacterial	*Staphylococcus aureus* inhibition with a minimum inhibitory concentration of 166 µg/ml	Shakir et al. (2019)
Leaves	–	Protein extract	Antibacterial	*S. aureus* inhibition with a minimum inhibitory concentration of 500 µg/ml	Shakir et al. (2019)
Leaves	–	Small proteins	Antibacterial	*Escherichia coli*, *S. aureus* and *Bacillus subtilis* inhibition with an inhibitory zone of 10 mm when in a dose of 0.80 mg/ml	Dahot (1998)
Seed	Naturally occurring peptide	Flo	Antibacterial	Reduced by around 50% the cell survival of *E. coli* after 2 h incubation	Fisch et al. (2004)
Flower	Naturally occurring peptide	Trypsin inhibitor	Antiprotozoal	Caused lysis of *Trypanosoma cruzi* with $LC_{50/24\,h}$ of 41.20 µg/ml	Pontual et al. (2009, 2018)
Flower	Naturally occurring peptide	Trypsin inhibitor	Antitermitic	Promoted the mortality of the termite *Nasutitermes corniger* workers	Pontual et al. (2009)

REFERENCES

Abdulkadir, A.R., D.D. Zawawi, and Md.S. Jahan. (2016) Proximate and phytochemical screening of different parts of *Moringa oleifera*. *Russian Agricultural Sciences* 42(1): 34–36.

Aderinola, T.A., T.N. Fagbemi, V.N. Enujiugha, A.M. Alashi, and R.E. Aluko. (2019). *In vitro* antihypertensive and antioxidative properties of trypsin-derived *Moringa oleifera* seed globulin hydrolyzate and its membrane fractions. *Food Science & Nutrition* 7(1): 132–138.

Aderinola, T.A., T.N. Fagbemi, V.N. Enujiugha, A.M. Alashi, and R.E. Aluko. (2018). Amino acid composition and antioxidant properties of *Moringa oleifera* seed protein isolate and enzymatic hydrolysates. *Heliyon* 4(10): e00877.

Ahn, J., M.-J. Cao, Y.Q. Yu, and J.R. Engen. (2013). Accessing the reproducibility and specificity of pepsin and other aspartic proteases. *Biochimica et Biophysica Acta (BBA) – Proteins and Proteomics* 1834(6): 1222–1229.

Ao, O., O.E. Ekun, T.I. David, O.E. Olorunfemi, and M.B. Oyewale. (2018). *Moringa oleifera* seed protein hydrolysates: kinetics of α-amylase inhibition and antioxidant potentials. *Global Advanced Research Journal of Medicine and Medical Sciences* 7(9), 190–201.

Apone, F., A. Barbulova, and G. Colucci. (2019). Plant and microalgae derived peptides are advantageously employed as bioactive compounds in cosmetics. *Frontiers in Plant Science* 10: 756.

Arise, A.K., R.O. Arise, M.O. Sanusi, O.T. Esan, and S.A. Oyeyinka. (2014). Effect of *Moringa oleifera* flower fortification on the nutritional quality and sensory properties of weaning food. *Croatian Journal of Food Science and Technology* 6(2): 65–71.

Arshad, Z.I.M., A. Amid, F. Yusof, I. Jaswir, K. Ahmad, and S.P. Loke. (2014). Bromelain: an overview of industrial application and purification strategies. *Applied Microbiology and Biotechnology* 98(17): 7283–7297.

Avilés-Gaxiola, S., C. Chuck-Hernández, and S.O. Serna Saldivar. (2018). Inactivation methods of trypsin inhibitor in legumes: a review. *Journal of Food Science* 83(1): 17–29.

Belsito, D.V., R.A. Hill, C.D. Klaassen, D.C. Liebler, J.G. Marks Jr, R.C. Shank, et al. (2017). Safety assessment of plant-derived proteins and peptides as used in cosmetics. www.cir-safety.org/sites/default/files/pltpep062017tent.pdf.

Braide, W., C.R. Ibegbulem, S.A. Adeleye, E.E. Mike-Anosike, P.B. Lugbe, I.J. Akien Alli, et al. (2017). Microbiological and nutritional analysis of roots and seeds of Moringa oleifera. International Journal of Research in Pharmacy and Biosciences 4(12): 19–24.

Bridgemohan, P., R. Bridgemohan, and M. Mohamed. (2014). Chemical composition of a high protein animal supplement from *Moringa oleifera*. *African Journal of Food Science and Technology* 5(5): 125–128.

Bruno, S.F., T.G. Kudre, and N. Bhaskar. (2019). Effects of different pretreatments and proteases on recovery, umami taste compound contents and antioxidant potentials of *Labeo rohita* head protein hydrolysates. *Journal of Food Science and Technology* 56(4): 1966–1977.

Bustamante, M.Á., M.P. Fernández-Gil, I. Churruca, J. Miranda, A. Lasa, V. Navarro, et al. (2017). Evolution of gluten content in cereal-based gluten-free products: an overview from 1998 to 2016. *Nutrients* 9(1): 21.

Carabarin-Lima, A., M.C. González-Vázquez, O. Rodríguez-Morales, L. Baylón-Pacheco, J.L. Rosales-Encina, P.A. Reyes-López, et al. (2013). Chagas disease (American trypanosomiasis) in Mexico: an update. *Acta Tropica* 127(2): 126–135.

Chakrabarti, S., S. Guha, and K. Majumder. (2018). Food-derived bioactive peptides in human health: challenges and opportunities. *Nutrients* 10(11): 1738.

Choi, Y.J., S. Hur, B.-D. Choi, K. Konno, and J.W. Park. (2009). Enzymatic hydrolysis of recovered protein from frozen small croaker and functional properties of its hydrolysates. *Journal of Food Science* 74(1): C17–C24.

Dahot, M.U. (1998). Antimicrobial activity of small protein of *Moringa oleifera* leaves. *Journal of the Islamic Academy of Sciences* 11(1): 6.

Daud, N.A., A.S. Babji, and S.M. Yusop. (2013). Antioxidant activities of red tilapia (*Oreochromis niloticus*) protein hydrolysates as influenced by thermolysin and alcalase. *AIP Conference Proceedings* 1571: 687.

Dewey, K.G., Z. Yang, and E. Boy. (2009). Systematic review and meta-analysis of home fortification of complementary foods. *Maternal & Child Nutrition* 5(4): 283–321.

Fisch, F., M. Suarez, and N. Mermoud. (2004). Flo antibacterial peptide from the tropical tree *Moringa oleifera*: a template for novel antibacterial agents. *Travail de diploma, Universite De Lausanne, Lausanne* 12(1–2): 1–4.

Garza, N.G.G., J.A. Chuc Koyoc, J.A. Torres Castillo, E. Alejandro, G. Zambrano, D.B. Ancona, et al. (2017). (Biofunctional properties of bioactive peptide fractions from protein isolates of moringa seed (*Moringa oleifera*). *Journal of Food Science and Technology* 54(13): 4268–4276.

Gidamis, A.B., J.T. Panga, S.V. Sarwatt, B.E. Chove, and N.B. Shayo (2003). Nutrient and antinutrient contents in raw and cooked young leaves and immature pods of *Moringa oleifera*, Lam. *Ecology of Food and Nutrition* 42(6): 399–411.

Glover-Amengor, M., R. Aryeetey, E. Afari, and A. Nyarko. (2017). Micronutrient composition and acceptability of *Moringa oleifera* leaf-fortified dishes by children in Ada-East district, Ghana. *Food Science & Nutrition* 5(2): 317–323.

Hekmat, S., K. Morgan, M. Soltani, and R. Gough. (2015). Sensory evaluation of locally-grown fruit purees and inulin fibre on probiotic yogurt in Mwanza, Tanzania and the microbial analysis of probiotic yogurt fortified with *Moringa oleifera*. *Journal of Health, Population, and Nutrition* 33(1): 60.

Huang, Z., D. Kang, J. Li, C. Guo, and S. Peng. (2020). Study on the optimization of the enzymatic hydrolysis of antimicrobial protein from *Moringa oleifera* leaves by response surface method. *IOP Conference Series: Earth and Environmental Science* 446(3): 032057.

Ijarotimi, O.S., O.A. Adeoti, and O. Ariyo (2013). Comparative study on nutrient composition, phytochemical, and functional characteristics of raw, germinated, and fermented *Moringa oleifera* seed flour. *Food Science and Nutrition* 1(6): 452–463.

Jain, A., R. Subramanian, B. Manohar, and C. Radha. (2019). Preparation, characterization and functional properties of *Moringa oleifera* seed protein isolate. *Journal of Food Science and Technology* 56(4): 2093–2104.

Kini, S.G., K.H. Wong, W.L. Tan, T. Xiao, and J.P. Tam. (2017). Morintides: cargo-free chitin-binding peptides from *Moringa oleifera*. *BMC Plant Biology* 17(1): 68.

Liang, L., C. Wang, S. Li, X. Chu, and K. Sun. (2019). Nutritional compositions of Indian *Moringa oleifera* seed and antioxidant activity of its polypeptides. *Food Science & Nutrition* 7(5): 1754–1760.

Liang, L., S. Cai, M. Gao, X. Chu, X. Pan, K.-K. Gong, et al. (2020). Purification of antioxidant peptides of *Moringa oleifera* seeds and their protective effects on H_2O_2 oxidative damaged Chang liver cells. *Journal of Functional Foods* 64: 103698.

Lin, L., Q. Zhu, M. Zhao, K. Zhao, Y. Tian, and Y. Yang. (2019). Purification of peptide fraction with antioxidant activity from *Moringa oleifera* leaf hydrolysate and protective effect of its *in vitro* gastrointestinal digest on oxidatively damaged erythrocytes. *International Journal of Food Science & Technology* 54(1): 84–91.

Loizzo, M.R., and R. Tundis. (2019). Plant antioxidant for application in food and nutraceutical industries. *Antioxidants* 8(1): 453.

Madane, P., A.K. Das, M. Pateiro, P.K. Nanda, S. Bandyopadhyay, P. Jagtap, et al. (2019). Drumstick (*Moringa oleifera*) flower as an antioxidant dietary fibre in chicken meat nuggets. *Foods* 8(8): 307.

Moyo, B., P.J. Masika, A. Hugo, and V. Muchenje. (2011). Nutritional characterization of Moringa (*Moringa oleifera* Lam.) leaves. *African Journal of Biotechnology* 10(60): 12925–12933.

Mótyán, J., F. Tóth, and J. Tőzsér. (2013). Research applications of proteolytic enzymes in molecular biology. *Biomolecules* 3(4): 923–942.

Mune, M.A., E.C. Nyobe, C.B. Bassogog, and S.R. Minka. (2016). A comparison on the nutritional quality of proteins from *Moringa oleifera* leaves and seeds. *Cogent Food & Agriculture* 2(1): 1213618.

Olson, M.E., R.P. Sankaran, J.W. Fahey, M.A. Grusak, D. Odee, and W. Nouman. (2016). Leaf protein and mineral concentrations across the 'Miracle Tree' genus *Moringa*. *PLoS ONE* 11(7): e0159782.

Paula, P.C., D.O.B. Sousa, J.T.A. Oliveira, A.F.U. Carvalho, B.G.T. Alves, M.L. Pereira, D.F. Farias, et al. (2017). A protein isolate from *Moringa oleifera* leaves has hypoglycemic and antioxidant effects in alloxan-induced diabetic mice. *Molecules* 22(2): 271.

Pinto, V.E.F., and A. Patriarca. (2017). *Alternaria* species and their associated mycotoxins. In *Mycotoxigenic Fungi* (pp. 13–32). New York: Springer.

Pontual, E.V., A.F.S. Santos, P.M.G. Paiva, and L.C.B.B. Coelho. (2009). Purification, characterization and termiticidal activity of Moringa oleifera flower peptides. www.sbbq. iq.usp.br/arquivos/2009/cdlivro/resumos/R8110.pdf.

Pontual, E.V., D.F. Pires-Neto, K. Fraige, T.M.M. Higino, B.E.A. Carvalho, N.M.P. Alves, et al. (2018). A trypsin inhibitor from *Moringa oleifera* flower extract is cytotoxic to *Trypanosoma cruzi* with high selectivity over mammalian cells. *Natural Product Research* 32(24): 2940–2944.

Rajapakse, N., E. Mendis, H.-G. Byun, and S.-K. Kim. (2005). Purification and *in vitro* antioxidative effects of giant squid muscle peptides on free radical-mediated oxidative systems. *The Journal of Nutritional Biochemistry* 16(9): 562–569.

Rayaprolu, S.J., N.S. Hettiarachchy, R. Horax, G. Kumar-Phillips, R. Liyanage, J. Lay, et al. (2017). Purification and characterization of a peptide from soybean with cancer cell proliferation inhibition. *Journal of Food Biochemistry* 41(4): e12374.

Saa, R.W., E.N. Fombang, E.B. Ndjantou, and N.Y. Njintang. (2019). Treatments and uses of *Moringa oleifera* seeds in human nutrition: a review. *Food Science & Nutrition* 7(6): 1911–1919.

Shakir, A., S. Riaz, M. Afzal, B. Zeshan, and S. Batool. (2019). Isolation and partial characterization of antimicrobial proteins/peptides from *Moringa oleifera*. *Life Science Journal of Pakistan* 1(2): 10–16.

Shebek, K., A.B. Schantz, I. Sines, K. Lauser, S. Velegol, and M. Kumar. (2015). The flocculating cationic polypeptide from *Moringa oleifera* seeds damages bacterial cell membranes by causing membrane fusion. *Langmuir* 31(15): 4496–4502.

Su, B., and X. Chen. (2020). Current status and potential of *Moringa oleifera* leaf as an alternative protein source for animal feeds. *Frontiers in Veterinary Science* 7: 1–53.

Szabó, A., M. Ludwig, E. Hegyi, R. Szépeová, H. Witt, and M. Sahin-Tóth. (2015). Mesotrypsin signature mutation in a chymotrypsin C (CTRC) variant associated with chronic pancreatitis. *Journal of Biological Chemistry* 290(28): 17282–17292.

Verma, K.S., and R. Nigam. (2014). Nutritional assessment of different parts of *Moringa oleifera* Lamm collected from Central India. *Journal of Natural Products and Plant Resources* 4(1): 81–86.

Walmsley, S.J., P.A. Rudnick, Y. Liang, Q. Dong, S.E. Stein, and A.I. Nesvizhskii. (2013). Comprehensive analysis of protein digestion using six trypsins reveals the origin of trypsin as a significant source of variability in proteomics. *Journal of Proteome Research* 12(12): 5666–5680.

Wong, K.H., W.L. Tan, S.G. Kini, T. Xiao, A. Serra, S.K. Sze, et al. (2017). Vaccatides: antifungal glutamine-rich Hevein-like peptides from *Vaccaria hispanica*. *Frontiers in Plant Science* 8: 1100.

Yun, Y.-R., S.-J. Oh, M.-J. Lee, Y.-J. Choi, S.J. Park, M.-A. Lee, et al. (2020). Antioxidant activity and calcium bioaccessibility of *Moringa oleifera* leaf hydrolysate, as a potential calcium supplement in food. *Food Science and Biotechnology* 29: 1563–1571.

Zambrowicz, A., M. Timmer, A. Polanowski, G. Lubec, and T. Trziszka. (2013). Manufacturing of peptides exhibiting biological activity. *Amino Acids* 44(2): 315–320.

Zhu, J., J. Qin, Z. Shen, J.D. Kretlow, X. Wang, Z. Liu, et al. (2012). Dispase rapidly and effectively purifies Schwann cells from newborn mice and adult rats. *Neural Regeneration Research* 7(4): 256.

Zhuang, M., L. Lin, M. Zhao, Y. Dong, D. Sun-Waterhouse, H. Chen, et al. (2016). Sequence, taste and umami-enhancing effect of the peptides separated from soy sauce. *Food Chemistry* 206: 174–181.

8 Antioxidant Properties of *Moringa* Species

Laura A. Contreras-Angulo
Centro de Investigación en Alimentación y Desarrollo A.C.,
Carretera a Eldorado Km. 5.5, Col Campo El Diez, Culiacán,
Sinaloa, 80110 México.

Alexis Emus-Medina
Universidad Autónoma de Sinaloa, Facultad de Medicina
Veterinaria y Zootecnia, Blvd. San Ángel 3886, Mercado de
Abastos, San Benito, Culiacán 80260, Sinaloa, México.

Manuel Bernal-Millan
Centro de Investigación en Alimentación y Desarrollo A.C.,
Carretera a Eldorado Km. 5.5, Col Campo El Diez, Culiacán,
Sinaloa, 80110 México.

J. Basilio Heredia
Centro de Investigación en Alimentación y Desarrollo A.C.,
Carretera a Eldorado Km. 5.5, Col Campo El Diez, Culiacán,
Sinaloa, 80110 México.

Erick P. Gutiérrez-Grijalva
Cátedras CONACYT-Centro de Investigación en
Alimentación y Desarrollo A.C., Carretera a Eldorado Km.
5.5, Col Campo El Diez, Culiacán, Sinaloa, 80110 México.
Corresponding author: erick.gutierrez@ciad.mx

CONTENTS

DOI: 10.1201/9781003108863-8

8.1   INTRODUCTION

The *Moringa* genus belongs to the Moringaceae family and comprises 13 species, *M. drouhardii, M. hildebrandtii, M. stenopetala, M. ovalifolia, M. peregrina, M. concanensis, M. oleifera, M. longituba, M. ruspoliana, M. borziana, M. arborea, M. rivae,* and *M. pygmea* (Abd Rani et al. 2018; Palada 2019). Genus *Moringa* is distributed around the world from Somalia (Kenya and Ethiopia), Southwest Angola (Namibia), India, Madagascar, Arabia, Northeast Africa, Pakistan, Philippines, Cambodia, to America (Central, North, and South), Florida, the Caribbean, and the Pacific Islands (Abd Rani et al. 2018; Sarwar 2017). Since ancient times people have used plants as food and medicine; within the plant kingdom, genus *Moringa* is widely used for the treatment of different diseases such as skin infections, anxiety, asthma, wounds, fever, diarrhea, malaria, stomach pain, cold, and diabetes. Mainly the leaves and seeds are used; however, roots, gum, flowers, and stems also have medicinal properties (Abd Rani et al. 2018; Palada 2019).

These properties of *Moringa* species are functions of the different groups of secondary metabolites present such as phenolic acids (chlorogenic acid, ellagic acid, sinapic acid, *p*-coumaric acid, ferulic acid, gallic acid, syringic acid, etc.), flavonoids (quercetin, kaempferol, rhamnetin, luteolin, myricetin, epicatechin, genistein, rutin, etc.), procyanidins, tannins, saponins, and vitamins (vitamin C and A; Singh and Sharma 2020). These compounds can generate a protective potential against oxidative stress, which has been linked to different diseases (Yeum et al. 2010). Different investigations have been carried out to demonstrate the antioxidant capacity of *Moringa* species. It has been found that the different species contain phytochemicals with antioxidant properties in both *in vitro* and *in vivo* assays.

8.2   BOTANICAL DESCRIPTION OF *MORINGA* SPECIES

The *Moringa* genus is found in the Moringaceae family along with *Anoma* and *Hyperanthera*. It is well known as the "drumstick" or "horseradish" family (Abd Rani et al. 2018). The genus comprises 13 species of trees with tropical and subtropical dicotyledonous flowers. The reports show that most *Moringa* species originated in India and Africa, but they have been introduced to several countries in the tropics (Padayachee and Baijnath 2012). Many of these tropical and subtropical species are in danger of extinction, including *M. arborea, M. borziana, M. longituba, M. rivae, M. ruspoliana,* and *M. stenopetala* (Shahzad et al. 2013). Current research is limited in *M. stenopetala, M. concanensis, M. peregrina,* but *M. oleifera,* which is native to India, has been studied extensively, and as a result, the species has been cultivated all over the world, specifically in Asia, Latin America, Florida, the Caribbean, and the Pacific Islands (Abd Rani et al. 2018).

According to their trunk type, *Moringa* species can be categorized into three groups. *Moringa stenopetala, M. drouhardii, M. ovalifolia,* and *M. hildebrandtii* have swollen trunks that store water and are known as "bottle trees". Meanwhile, *M. peregrina, M. concanensis,* and *M. oleifera* have slender trunks known as "slender trees". The remaining species (*M. arborea, M. borziana, M. longituba, M. pygmaea, M. rivae,* and *M. ruspoliana*) are tuberous shrubs endemic to northeast Africa (Abd Rani et al. 2018; Olson 2019).

8.2.1 SPECIES GROUP "BOTTLE TREES"

Moringa stenopetala is a native tree of Southern Ethiopia, Northern Kenya, and Eastern Somalia, and is therefore known as African moringa (Gebregiorgis et al. 2012). *M. stenopetala* is a huge tree that often branches near the base. The leaf glands of this species are among the most active of the genus, often causing the leaves to glow with droplets of clear, sticky exudate that attracts ants. *Moringa stenopetala* has attracted interest because its larger size appears to make it more resistant to drought than *M. oleifera* (Olson 2019). *Moringa stenopetala* is a tree, up to 10 m tall, with a swollen, bottle-shaped trunk, whitish bark, and pinnate leaves with elliptical to oval leaflets. The bisexual flowers are regular and have free sepals and cream-colored pink petals. The fruits are elongated capsules with three valves 20–50 cm long, and the seeds are elliptical to trigonous and 6–9 cm long with three thin wings (Padayachee and Baijnath 2012).

Moringa drouhardii is a small tree endemic to the Toliara province in southwestern Madagascar, and this species usually has a single fat trunk, 2–18 m high, whitish bark, have 3-pinnate leaves with long, slender, and fruity leaflets with shrinkage between the seeds (2–3 cm long), pitted like golf balls and without wings. The flowers are bisexual, regular, and ovate with yellowish-white petals. *M. drouhardii* grows very quickly and can reach flowering in just two years. The species has been documented in southwest and southeast sites of the island, but the extent to which the tree is found between these two areas has not been studied in detail (Padayachee and Baijnath 2012; Olson 2019).

Moringa ovalifolia, known as ghost or ghost tree, is the only species in the genus *Moringa* that is endemic from central Namibia to southern Angola (Olson 2019; Hausiku et al. 2020). It is an erect, deciduous tree that grows up to 7 m tall, with smooth, resinous bark, and alternately arranged compound leaves with leaflets that grow up to 25 mm long. It has white flowers that grow up in branched axillary sprays with 4–5 petals and brown, three-angled pendulous pods 400 mm long (Padayachee and Baijnath 2012). Mountainous slopes characterize their habitats, and they grow interspersed with boulders (Hausiku et al. 2020).

Moringa hildebrandtii is an endemic Madagascan tree (G. Kumar et al. 2019), which can grow up to 25 m tall; it is similar to *M. drouhardii* but with larger, thicker leaves that generally have red on the trunks between the leaflets, along with a rough brown bark that exudes a reddish gum and flowers that are slightly bilaterally symmetric (Olson 2019). The fruits are spindle-shaped capsules between 450 and 650 mm long, and they contract between the seeds, which are pale brown, ovoid trigonous, with winged edges and 35–40 mm long (Padayachee and Baijnath 2012).

8.2.2 SPECIES GROUP "SLENDER TREES"

Moringa oleifera is the most known and cultivated of the 13 species of trees and shrubs of the genus *Moringa* (Padayachee and Baijnath 2012; Gandji et al. 2018). *M. oleifera* is a small tree native to the sub-Himalayan regions of northwestern India, and is now native to many regions of Africa, Arabia, Southeast Asia, the Pacific Islands, the Caribbean, and South America (Razis et al., 2014). It is a deciduous or evergreen fast-growing tree, 10–12 m tall. The leaves are bipinnate or, more commonly, tripinnate, up to 45 cm long, alternate, and spirally arranged on twigs. The flowers are fragrant and bisexual, surrounded by five uneven, yellowish-white, finely streaked petals (Chukwuebuka 2015). It is identified by the long and woody pod shape of the fruit, and when ripe, it opens into three valves that contain seeds with longitudinal wings (Velázquez-Zavala et al. 2016). Each pod usually contains up to 26 seeds; they are dark green during development and take approximately 3 months to mature after flowering. When injured, the bark exudes gum, which is initially white, but changes to reddish-brown or brownish-black on exposure (Chukwuebuka 2015). *M. oleifera* is relatively easily spread by both sexual and asexual means. It has a low demand for water and nutrients from the soil, facilitating its production and management. They are resistant to drought and can withstand a wide range of rainfall and soil conditions. Therefore, they are available throughout the year (Sahay 2017).

Moringa peregrina is an extremely fast-growing tree or shrub that commonly reaches around 3–10 m in height only 10 months after planting the seed (Said-Al Ahl et al. 2017). It is geographically distributed from tropical Africa to eastern India (El-Hak et al. 2018). It has a grayish-green bark adapted to high aridity. The leaves are 30–40 cm long, alternate, ovate and deciduous. One of the unique characteristics of *M. peregrina* is the loss of its leaflets when the leaves mature, leaving the rachis of the leaves bare (Senthilkumar et al. 2018). The flowers (10–15 mm long) are generally yellowish-white to pink, bisexual, and harbor insect pollination characteristics, with large, showy, slightly scented, and zygomorphic petals (Robiansyah et al. 2014). The seeds are wingless, triangular, and 25 mm long and 10–12 mm in diameter, they have a short germination time and a high seedling growth rate (Salaheldeen et al. 2014; Asghari et al. 2015).

Moringa concanensis is the closest living relative of *M. oleifera* and is still found in the wild in fragments of dry tropical forest from eastern Pakistan to India's southern tip and in a small part of Bangladesh (Olson 2019). *M. concanensis* is a small tree that is glabrous except for the tender leaves and inflorescences; the central trunk is covered with an extremely distinctive layer of heavily furrowed bark (Padayachee and Baijnath 2012). Its leaves are perennial with a spreading crown, up to 7–8 feet, alternate, bipinnate, and oval. Its flowers are large, white, hermaphrodite, irregular axillary panicles with yellow petals, streaked with red, and oblong (Santhi and Sengottuvel 2016). The capsules are straight, triangular, and slightly narrow between the seeds, white or pale yellow, and have three angles (Balamurugan and Balakrishnan 2013).

8.2.3 SPECIES GROUP "TUBEROUS SHRUBS"

Moringa arborea is one of the rarest moringas and inhabits rocky canyons in limestone lowlands. It is a shrub or tree distributed on the border between Kenya and Ethiopia (Padayachee and Baijnath 2012; Olson 2019). It is characterized by its large flowers, seeds, and glabrous trunks, and it has smooth, gray, thin bark; it also has large sprays of pale pink and wine-red flowers. A striking feature of *M. arborea* is its very long fruits, like a 60-cm long green bean (Padayachee and Baijnath 2012; Arora et al. 2013; Olson 2019).

Moringa borziana is a woody herb or small shrub generally found in disturbed grasslands or shrublands. It is distributed from southern Kenya to the Kisumayu region of southern Somalia, generally within 100 miles of the coast (Padayachee and Baijnath 2012; Olson 2019). It bears one or two trunks that die back to the tuber every few years, sometimes due to variation of environmental parameters the plant grows into a small tree; it has greenish cream to yellow with brown flowers (Arora et al. 2013; Olson 2019). The leaves are pinnate, glabrous, and stipitate; the leaflets are pale green to yellow and elliptical to ovate. The fruits are purplish-brown with waxy flowers, and the seeds are 3.8 cm long in total with three striking wings (Padayachee and Baijnath 2012).

Moringa longituba is a distinctive species of the *Moringa* genus. It is a shrub or subshrub inhabiting low hillsides in dry scrub on rocky soil or wooded grasslands on deep soils. It is distributed in northeastern Kenya, southeastern Ethiopia, and a large part of Somalia. It is 2–6 m tall and has a smooth, pale gray bark. No other species has bright red flowers or the bases of petals and sepals fused to form a long tube. It has a large tuber deep underground, usually with a small shoot that reaches knee height above the ground. The leaves are 2–3 pinnate, densely pubescent when young and later glabrescent, and the leaflets are oblong, elliptical, or ovate. The fruits are purple with flowers, and the seeds are 2–3 cm long and have wings (Padayachee and Baijnath 2012; Arora et al. 2013; Olson 2019).

Moringa pygmaea is an extremely rare, delicate, and geographically restricted species, a shrub or herb, which is mainly distributed in northern Somalia. The aerial part of the plant is only a few centimeters high and has several immature leaves, confirming that it is the smallest *Moringa*. The leaves are pinnate, glabrous with 3–4 pairs of pinnae and tiny leaflets. The flowers are bisexual, borne in axillary panicles, and yellow. The fruits have a capsules form, ribbed, and sometimes appear with an elongated beak, and their seeds are three-winged or wingless and without endosperm (Padayachee and Baijnath 2012; Arora et al. 2013; Olson 2019).

Moringa rivae is a variable species, and it is usually a shrub or a small tree. It has many similar characteristics with *M. borziana* and *M. arborea*. It is distributed in various parts of Kenya or southern Ethiopia, finding it on rocky hillsides in high thickets. Its leaves are alternate, pinnate, and the leaflets are oblong to elliptical and glabrous. Its flowers are pale cream with brown spots at the tips. The seeds are embedded in the valve holes and have three angles, blackish and rounded wings (Padayachee and Baijnath 2012; Arora et al. 2013; Olson 2019; Saleem et al. 2020b).

Moringa ruspoliana is a small tree with large, hard, thick leaves; large flowers; and the thick main root that becomes more globose as the plant ages. It inhabits altered grass forests on low limestone plateaus and is distributed in Somalia, Ethiopia, and northern Kenya. Besides *M. longituba*, it is the only other species with red flowers. These flowers are striped pinkish red with green at the base and are the largest of the family, up to 3 cm long. The leaves of *M. ruspoliana* are distinguished for having enormous leaflets up to 15 cm wide and being pinnate only once, in contrast to all other members of the genus (Padayachee and Baijnath 2012; Arora et al. 2013; Olson 2019).

8.3 BIOACTIVE COMPOUNDS FROM *MORINGA* SPECIES

It is recognized that the consumption of fruits and vegetables provides a significant amount of nutrients and bioactive compounds. Vegetables have aroused great interest because despite not having a defined nutritional function, they have been related to the prevention of various diseases; therefore, consumption of moringa has been discussed as new public health recommendations. Bioactive or phytochemical compounds are present various structures and can be divided into nitrogenous, sulfurous, terpenoid, and phenolic substances (Liu 2013; Martínez-Navarrete et al. 2008; Weaver 2014).

The moringa tree is recognized worldwide and has been used for multiple purposes, mainly for consumption due to its important nutritional contribution of protein, minerals, and antioxidants; for this reason, it is used in some countries to combat malnutrition (Leone et al. 2015). Also, it has been described that the different parts of the tree, mainly the leaves, have a significant amount of bioactive compounds such as phenolic acids, flavonoids, glucosinolates, alkaloids, and carotenoids, which have demonstrated pharmacological properties such as antioxidant, antidiabetic, anticancer and anti-inflammatory (Ma et al. 2020; Chhikara et al. 2021). In this context, it is important to know the different types of bioactive compounds characteristic of the different moringa species that could indicate their possible medicinal applications.

8.3.1 PHENOLIC ACIDS

Phenolic acids are among the most important groups of bioactive compounds in various plant species. They are derived from the secondary metabolism of the shikimic acid and phenylpropanoid pathways; they impart some characteristics to plants, such as taste and smell. Phenolic acids have been shown to have anti-inflammatory, anti-atherogenic, antimicrobial, antithrombotic, antidiabetic, and cardioprotective properties. The phenolic acid structure is made up of a phenolic ring and at least one organic carboxylic acid function. They are divided into two groups: hydroxybenzoic acid, with structure C6–C1, and hydroxycinnamic acid, with structure C6–C3 (Goleniowski et al. 2013; Heleno et al. 2015; N. Kumar and Goel 2019; Rashmi and Negi 2020).

The moringa tree has been classified as a plant rich in phenolic compounds, highlighting the presence of phenolic acids in different species. For example, in *M. ovalifolia* the presence of *cis* and *trans* isomers of 3-acyl, 4-acyl, and 5-acyl *p*-coumaroylquinic, caffeoylquinic, and feruloylquinic acids has been reported, in addition to 3,5-di-caffeoylquinic acid, 3-caffeoylquinic-glycoside, and two isomers of the glycosides of 4-caffeoylquinic acid (Makita et al. 2017). Similarly, Juhaimi et al. (2017) analyzed *M. peregrina* leaves and identified different phenolic acids such as gallic acid (0.930 mg/100 g dry weight [DW]), protocatechuic acid (0.070 mg/100 g DW), 4-hydroxybenzoic acid (0.740 mg/100 g DW), caffeic acid (0.250 mg/100 g DW), syringic acid (0.08 mg/100 g DW), *p*-coumaric acid (0.140 mg/100 g DW), chlorogenic acid (0.030 mg/100 g DW), *trans* ferulic acid (0.19 mg/100 g DW), and *trans*-cinnamic acid (0.270 mg/100 g DW).

On the other hand, *M. oleifera* is perhaps the most studied species, and different phenolic acids have been identified in this species. For instance, Álvarez-Román et al. (2020) found phloretic acid, caffeic acid, quinic acid, caffeic acid *O*-glucoside, and chlorogenic acid in hydroalcoholic extracts of leaves. Likewise, Saleem et al. (2020a) analyzed different extracts of *M. oleifera* leaves (methanolic, hexanic, ethyl acetate, butanol, and aqueous) and identified gallic acid, caffeic acid, benzoic acid, sinapic acid, syringic acid, coumaric acid, vanillic acid, ferulic acid, and chlorogenic acid. Consistent results were found by Oguntibeju et al. (2020), who report the presence of chlorogenic acid (250 µg/ml), caffeic acid (8119 µg/ml), and coumaric acid (15740 µg/ml) in leaves of *M. oleifera*. Asgari-Kafrani et al. (2020) found a caffeic acid content of 0.348 mg/100 g in leaves and 0.294 mg/100 g in stems, as well as gallic acid with a concentration of 0.212 mg/100 g in leaves and 0.170 mg/100 g in stems.

In another study conducted by Juhaimi et al. (2017), gallic acid (1.145 mg/100 g DW), protocatechuic acid (0.051 mg/100 g DW), 4-hydroxybenzoic acid (0.710 mg/100 g DW), caffeic acid (0.231 mg/100 g DW), syringic acid (0.06 mg/100 g DW), *p*-coumaric acid (0.119 mg/100 g DW), chlorogenic acid (0.028 mg/100 g DW), *trans* ferulic acid (0.235 mg/100 g DW), and *trans*-cinnamic acid (0.299 mg/100 g DW) were identified. In the case of *M. oleifera* seeds, phenolic acids and derivatives such as 3,4-dihydroxycinnamic acid, echinacoside, vanillic acid-4-O-β-D-glucopyranoside, and protocatechuic aldehyde, among others, have been identified (Gu et al. 2020).

8.3.2 FLAVONOIDS

Flavonoids are a large group of secondary metabolites belonging to phenolic compounds and synthesized through the phenylpropanoid pathway. They are widely distributed in the plant kingdom, found in all parts of plants. More than 6500 flavonoids have been identified, and they are responsible for color and aroma in flowers; act as attractants for pollinators; participate in the transport of auxin; protect plants against different types of stress; act as protectors against UV, insects, and microorganisms; and modulate the levels of reactive oxygen species, among other things (Buer et al. 2010; Samanta et al. 2011; Kumar and Pandey 2013). The basic structure of flavonoids consists of 15 carbons organized in a C6–C3–C6 form conformed by two benzene rings linked by a heterocyclic pyran ring. Flavonoids are commonly divided into flavones, flavonols, flavanones, flavanonols, flavanols or catechins, anthocyanins, and chalcones. The consumption of some flavonoids has been linked to health benefits due to their antioxidant, anti-inflammatory, antimutagenic, and anticancer properties (Xiao et al. 2011; Kumar and Pandey 2013; Panche et al. 2016).

The high flavonoid content characterizes the moringa tree in its different species, and several researchers have managed to identify an important variety of these metabolites. The methanolic extracts of *M. peregrina* leaves showed a total flavonoid content of 33.40 mg of quercetin/g extract, and a lower amount was found in aqueous extracts with 13.89 mg of quercetin/g extract (El-Awady et al. 2016). In the same way, Al-Dabbas (2017) researched with different extracts (methanol, ethyl acetate, and hexane) of *M. peregrina* and found that the ethyl acetate extracts had the highest content of total flavonoids, with 194.3 mg of rutin/g in leaves; while in seeds the majority content was observed with the hexane extracts (214.2 mg of rutin/g).

In a different study conducted in hydroalcoholic leaf extracts of *M. ovalifolia*, the presence of quercetin, kaempferol rutinoside, and isorhamnetin, all glycosylated with rutinoside, were identified (Makita et al. 2016). In other moringa species, such as *M. peregrina*, *M. stenopetala*, and *M. oleifera*, 11 flavonoids were identified, such as naringin, rutin, hesperidin, rosmarinic, quercetrin, quercetin, naringenin, kaempferol, hispertin, apigenin, and OH flavone-7, highlighting the content of kaempferol (8410.12 mg/l) in *M. stenopetala*, rutin (1200.06 mg/l) in *M. peregrina*, and hesperidin (3846.97 mg/l) in *M. oleifera* (Abo El-Fadl et al. 2020). Different parts of *M. olifera* were studied (leaves, seeds, stem, root) with different extracts (water, 50% ethanol, 70% ethanol, and 90% ethanol), showing a total flavonoid content between 2.7 and 123 mg of rutin/g; the highest flavonoid content corresponded to the leaf, followed by the stem, root, and seed (F. Wang et al. 2020).

On the other hand, Oguntibeju et al. (2020) found quercetin (17.03 µg/ml), rutin (9.55 µg/ml), and myricetin (108.02 µg/ml) in leaves of *M. oleifera*, while Asgari-Kafrani et al. (2020) found quercetin in leaves and stems at concentrations of 0.045 and 0.444 mg/100 g DW. Furthermore, Verma et al. (2009) identified the presence of flavonoids such as kaempferol (497.6 µg/g), quercetin (807 µg/g), and rutin (190 µg/g) in ethyl acetate leaf extract fractions. Also in hydroalcoholic leaf extracts the presence of apigenin glucoside and quercetin-*O*-glucoside have been reported (Álvarez-Román et al. 2020). In addition, they have identified in seeds of *M. oleifera* the presence of flavonoid derivatives such as 5,7-dimethoxy-4'-hydroxyflavone-4'-O-α-L-rhamnose(1-2)-β-D-glucoside, nevadensin-7-O-[α-L-rhamnosyl(1-6)]-β-D-glucoside, 4',5,6,7-tetramethoxy-flavone, taxifolin-3-*O*-glucoside, cyanidin-3-glucoside, and catechin-7-*O*-β-D-glucopyranoside, among others (Gu et al. 2020).

8.3.3 CAROTENOIDS

Carotenoids are terpenoid pigments synthesized by plants, fungi, and bacteria, and more than 600 carotenoids are known in nature. In plants they provide characteristic colors such as yellows, oranges, and reds, with their main role in photosynthesis and photoprotection, as well as in the biosynthesis of phytohormones, among other functions. Carotenoids are derived from eight isoprene units with an extensive conjugated double-bond system. They are divided into two large groups: xanthophylls that contain oxygen, and carotenes that are hydrocarbons. Various investigations have linked the consumption of carotenoids in the prevention of certain diseases such as cardiovascular, cancer, and diabetes mellitus, among others (Roohbakhsh et al. 2017; Ribeiro et al. 2018; Khalid et al. 2019; von Lintig et al. 2019; Rowles and Erdman Jr 2020).

The medicinal properties of the moringa tree are attributed to its content of bioactive compounds, including carotenoids, which have been reported in some species. For example, in a study that consisted of the foliar application of amino acids (1, 2 and 3 cm³/l) and extracts of yeast (1, 2, and 3 g/l) in *M. oleifera* and *M. ovalifolia*, a content of total carotenoids in leaves between 7.96–10.55 and 7.70–9.21 mg/g fresh weight (FW), respectively, was found (El-baset 2017). In another species, *M. concanensis*, a total carotenoid content between 25.93 and 70.64 mg/100 g has been reported (Anitha et al. 2018). In fresh *M. oleifera* leaves, some carotenoids have been identified and quantified such as *trans*-luteoxanthin (5.2 mg/100 g), 13-*cis*-lutein (2.31 mg/100 g), *trans*-lutein (36.88 mg/100 g), *trans*-zeaxanthin (5.46 mg/100 g), 15-*cis*-β-carotene (0.69 mg/100 g), and *trans*-β-carotene (18.27 mg/100 g) (Saini et al. 2014b).

Saini et al. (2014a) carried out a study on fresh leaves of eight cultivars of *M. oleifera*. They found the carotenoids all-*E*-luteoxanthin (2.58–5.68 mg/100 g), 13-Z-lutein (1.58–7.12 mg/100 g), all-*E*-lutein (17.60–41.16 mg/100 g), all-*E*-zeaxanthin (2.26–13.54 mg/100 g), 15-Z-β-carotene (0.40–0.69 mg/100 g), and all-*E*-β-carotene (11.86–23.15 mg/100 g). In addition, all the above mentioned carotenoids could be identified in fruit and flowers for the PKM-1 cultivar with a total content of carotenoids of 29.66 and 5.44 mg/100 g, respectively. Therefore, the leaf is the part of the tree with the highest content, being between 44.30 and 80.48 mg/100 g for the eight cultivars analyzed. Consistent results are reported by Teixeira et al. (2014), who found β-carotene and lutein as the major carotenoids in *M. olefeira* leaves with concentrations of 161 and 47 µg/g DW. In the case of *M. oleifera* seeds, a total carotenoid content of between 13.53 and 19.56 mg/100 g DW has been found (Barakat and Ghazal 2016). In the same way, some research on oils obtained from *M. oleifera* has been reported to contain carotenoids. Dinesha et al. (2018) investigated different oil extraction processes with supercritical CO_2, Soxhlet, and solvent and found a total carotenoid content of 17.06, 15.72, and 15.26 ppm, respectively. Also, the moringa oil refining process was analyzed and a decrease in carotenoids was found during all steps; crude oil (6.5 mg/g), neutralized oil (2.0 mg/g), degummed oil (3.1 mg/g), and bleaching oil (1.2 mg/g) (Sánchez-Machado et al. 2015).

8.3.4 GLUCOSINOLATES

Glucosinolates are compounds derived from the secondary metabolism of plants, present in 13 botanical families but mainly in the Brassicaceae family. Its general structure is made up of a β-D-thioglucose group, a sulfonated oxime group, and a side chain derived from amino acids. They are divided into three classes depending on their precursor amino acid as aliphatics glucosinolates, indole glucosinolates, and aromatic glucosinolates. They act mainly as defense compounds in plants and both glucosinolates and their derivatives (isothiocyanates, thiocyanates and nitriles) have been studied as antioxidant, antidiabetic, antimicrobial and anticancer compounds (Bohinc et al. 2012; Dinkova-Kostova and Kostov 2012; Capuano et al. 2017; Sánchez-Pujante et al. 2017; Bell et al. 2018). These compounds have been found in the moringa tree and their presence is related to its medicinal properties attributed to its consumption; in some investigations in different species it has been possible to identify some glucosinolates.

Maldini et al. (2014) identified and quantified glucosinolates in different parts of *M. oleifera* seedlings (pulp seed, seed coat, leaves, and roots), observing glucomoringin as the majority compound in all parts analyzed, with the pulp having the highest content (8619.44 mg/100 g), followed by leaves (77.7 mg/100 g), seed coat (28.27 mg/100 g), and roots (3.99 mg/100 g). In addition, it was possible to identify other glucosinolates such as 3-hydroxy-4-(α-L-rhamnopyranosyloxy) benzyl glucosinolate, 4-(2'-*O*-acetyl-α-L-rhamnopyranosyloxy) benzyl glucosinolate, 4-(3'-*O*-acetyl-α-L-rhamnopyranosyloxy) benzyl glucosinolate, 4-(4'-*O*-acetyl-α-L-rhamnopyranosyloxy) benzyl glucosinolate, glucotropaeolin, glucosibalbin, glucoraphanin, and glucoiberin.

Different extracts of *M. oleifera* leaves (aqueous extract, methanolic extract, glucosinolate-rich extract, and glucosinolate-rich hydrolyzed extract) showed the highest concentrations of different glucosinolates, such as glucomorin (24.51

mg/g), acetate glucomorin A (1.11 mg/g), glucomorin b acetate (5.81 mg/g), and glucomorin c acetate (24.06 mg/g), with the glucosinolate-rich extract (Cuellar-Núñez et al. 2020). Similarly, Bennett et al. (2003) analyzed the glucosinolate content of *M. oleifera* and *M. stenopetala* in different parts of the plant, and found that 4-(α-L-rhamnopyranosyloxy)-benzyl glucosinolate was the only glucocinolate in the seeds at concentrations of 201 and 256 mg/g DW, respectively. Also, in both young and old leaves, 4-(α-L-rhamnopyranosyloxy)-benzyl glucosinolate was found for both species and three monoacyl isomeres of this glucosinolate. In the case of stem and root in *M. oleifera*, 4-(α-L-rhamnopyranosyloxy)-benzyl glucosinolate was found at concentrations of 16.3 and 20.4 mg/g DW, respectively, while in *M. stenopetala* the concentrations were 13.6 and 40.9 mg/g DW for the stem and root, respectively. Furthermore, benzyl-glucosinolate was found in roots in high concentrations with 22.7 and 30.8 mg/g DW for *M. oleifera* and *M. stenopetala*, respectively. A study with *M. stenopetala* showed glucoconringiin in complete seeds (21.66 mg/g DW), seeds without testa (30.16 mg/g DW), and seed testa (0.28 mg/g DW); while *O*-(rhamopyranosyloxy)-benzyl-glucosinolate was found in higher concentrations in leaves (5.70 mg/g DW), roots (1.57 mg/g DW), complete seeds (129.57 mg/g DW), seeds without testa (187.49 mg/g DW), and seed testa (1.42 mg/g DW) (Mekonnen and Dräger 2003)

8.4 ANTIOXIDANT PROPERTIES OF *MORINGA* SPECIES

The bioactive compounds present in *Moringa* species have shown the ability to neutralize free radicals (reactive oxygen species [ROS] and reactive nitrogen species [RNS]). The disbalance between the free radicals and antioxidants in the cells generate the oxidative stress, the main cause of the appearance of diseases, because it induces oxidative alterations in macromolecules (Bullon et al. 2009; Cerdá et al. 2014; Cabello-Verrugio et al. 2018). Antioxidants can reduce oxidative stress by providing a natural defense to the cell and by acting as free radical scavengers (Verma et al. 2009). On the other hand, the regulation of the ROS can be by the action of enzymatic and non-enzymatic antioxidants (antioxidant compounds). Superoxide dismutase (SOD: eliminates the superoxide anion), catalase, and glutathione peroxidase (GPx: reduces the hydrogen peroxide) are the main enzyme antioxidants (Birben et al. 2012; Veskoukis et al. 2012).

8.4.1 IN VITRO STUDIES

As has been previously mentioned, *Moringa* plants are the source of bioactive compounds such as phenolic compounds, terpenes, and some alkaloids, and a potential source of bioactive peptides. Recent studies have shown a correlation between the consumption of antioxidant molecules and decreased risk and delayed onset of non-communicable diseases. Before we begin on the *in vitro* studies of the *Moringa* antioxidant potential it is important to define what an antioxidant is. Currently, the most accepted term is the one stated by Halliwell and Gutteridge (2015): "an antioxidant is any substance that, when present at low concentrations compared with those of an oxidizable substrate, significantly delays or prevents oxidation of that substrate". Most phytochemicals, like phenolic compounds and phenolic terpenes, fit this description. However, the mode of action of antioxidants is evaluated based on the chemical properties and oxidative–reduction relationship between the antioxidant molecule and the free radical. Granato et al. (2018) mentions that antioxidants can be categorized based on their:

- Function. The evaluation methods focus on the ability of the antioxidant to scavenge free radicals, non-radical oxidizing agents, transition metal-chelating agents, or if they stimulate endogenous antioxidant molecules.
- Polarity. There are methods specific for lipophilic and hydrophilic antioxidants.
- Source. Methods focusing on the evaluation of endogenous or exogenous antioxidants.
- Mechanism. Based on their electrophile or nucleophile chemical abilities, or most commonly known as hydrogen atom transfer (HAT) and single electron transfer (SET) mechanisms.

The most commonly used methods for evaluation of the antioxidant capacity of natural antioxidants like phytochemicals from *Moringa* are the HAT assays like oxygen radical absorbance capacity (ORAC), inhibition of lipoperoxidation, β-carotene bleaching assay; and the SET assays like the cupric-ion reducing antioxidant capacity (CUPRAC), the ferric reducing ability of plasma (FRAP), the Folin–Ciocalteu reagent, and the scavenging capacity of the 1,1-diphenyl-2-picrylhydrazyl (DPPH), among others (Granato et al. 2018). These *in vitro* assays have several advantages: they are cost- and time-effective, and provide an early insight of the potential free-radical scavenging capacity of molecules. However, they fail to translate into *in vivo* conditions. Thus, results from these assays must be interpreted carefully to avoid out-of-place statements (Fraga et al. 2014; Granato et al. 2018).

The antioxidant properties of *Moringa* extracts have been focused mostly on its content and distribution of phenolic compounds. Some of the phenolic compounds identified in *Moringa* are quercetin derivatives, kaempferol derivatives, chlorogenic acid, and caffeoylquinic acid isomers. Most studies are focused on aqueous or alcoholic extracts of *Moringa oleifera* leaves, because this is the most common plant part traditionally used. However, other studies have also shown that moringa fruits, roots, and seeds are also an important source of phytochemicals with antioxidant properties (Godinez-Oviedo et al. 2016; Abd Rani et al. 2018; Ahl et al. 2017). In this sense, Table 8.1 shows recent studies of the *in vitro* evaluation of the antioxidant potential of moringa plants.

It is important to mention that *in vitro* antioxidant assays are quick and cost-effective ways to evaluate the reducing, singlet oxygen quenching, and hydrogen donor capacities of the bioactive molecules in a sample using different free radicals. Thus, in the aforementioned studies it was confirmed that bioactive molecules obtained from moringa plants are potential antioxidant agents. However, this must lead to more cellular and *in vivo* studies to prove the antioxidant effect and mechanism of action of the antioxidant molecules of moringa. Moreover, before *in vivo*

TABLE 8.1

Summary of *in vitro* antioxidant reports of *Moringa* plants

Moringa source	Plant part	Methods	Description	Compounds associated	Reference
Moringa oleifera Lam.	Leaves, stem, root, bark	Hypoxanthine/xanthine oxidase assay system 2-deoxyguanosine assay	By the xanthine oxidase method, all extracts showed antioxidant activity. Extracts from roots, leaves, and stem bark showed IC_{50} values of 16, 30, and 38 µg/ml, respectively. The antioxidant capacity by the 2-deoxyguanosine assay showed IC_{50} values of 40, 58, and 72 µg/ml for extracts from leaves, stem bark, and roots, respectively	Chlorogenic acid, rutin, kaempferol glucoside	Atawodi et al. (2010)
M. oleifera	Leaves	Folin Ciocalteu's method DPPH Radical scavenging The authors tested the effect of different extraction methods: maceration with shaking, ultrasound-assisted, and Soxhlet extraction	The authors report a significant effect ($P \leq 0.05$) of the extraction technique and the extraction time on the antioxidant capacity and reducing capacity of the extracts. Folin–Ciocalteu's values ranging between 66.9 and 76.3 mg GAE/g, and DPPH inhibition IC_{50} values of 388.1–433.5 µg/cm³	α-Tocopherol, β-sitosterol, fucosterol	Dzieciol (2020)

Species	Plant part	Assay	Findings	Antioxidant compounds	References
M. oleifera	Leaves	FRAP, ABTS, DPPH, and ORAC assays	The authors showed that *M. oleifera* grown in Spain has higher antioxidant capacity than 28 vegetables commonly sold in local markets in Spain. Antioxidant values of *M. oleifera* samples were 6208 µmol TE/100 g, 3055 µmol TE/100 g, 3962 µmol TE/100 g, and 10805 µmol TE/100 g in fresh weight moringa by the TEAC, DPPH, FRAP, and ORAC assays, respectively	β-Carotene, chlorophyll a, chlorophyll b	Gonzalez-Romero et al. (2020)
M. oleifera	Seed flour	DPPH and FRAP assays	Free and bound phenolics of *M. oleifera* seeds were evaluated. Free phenolics showed higher antioxidant capacity with DPPH IC_{50} values of 14.9 mg/ml. However, in FRAP the bound phenolics showed higher values of 5.62 g/100 g	Gallic acid, epicatechin, caffeic acid	R. S. G. Singh et al. (2013)
M. oleifera	Leaves	Copper reduction capacity method	Methanol extracts showed higher antioxidant capacity than the aqueous ones; with antioxidant values of 160.19 and 9.18 nmol TE/µl, respectively	Not reported	Landazuri et al. (2020)
M. oleifera	Leaves and fruits	$FeCl_3$ reducing power, FRAP, DPPH, and the total antioxidant capacity estimation assays	There was a concentration- and dose-dependent effect. The highest antioxidant capacity values were obtained at 1000 µg/ml	Not reported	Luqman et al. (2012)

(continued)

TABLE 8.1 (Continued)
Summary of in vitro antioxidant reports of Moringa plants

Moringa source	Plant part	Methods	Description	Compounds associated	Reference
M. oleifera	Leaves	ABTS, DPPH, and nitric oxide scavenging activity assays	Acetone *M. oleifera* extracts had higher antioxidant activity than the aqueous ones. At a concentration of 1 mg/ml, the extracts showed antioxidant values of 95.27%, 98.24%, and 98.47% of inhibition of the ABTS, DPPH and nitric oxide radicals, respectively	Not reported	Moyo et al. (2012)
M. oleifera	Leaves from 13 *M. oleifera* cultivars	DPPH, FeCl$_3$ reducing power, and the β-carotene-linoleic acid model system assays	The antioxidant activity of *M. oleifera* depended on the cultivar. Cultivars from the World Vegetable Centre and Thailand were the best with DPPH inhibitory IC$_{50}$ values from 16.40 to 26.88 μg/mL. The highest FeCl$_3$ reducing power and β-carotene bleaching highest values were from cultivars from the World Vegetable Centre	Not reported	Ndhlala et al. (2014)
M. oleifera	Leaves and flowers	DPPH and reducing power assays	The authors showed that extracts from moringa flowers and leaves had higher antioxidant capacity than commonly used vegetables like broccoli, spinach, cauliflower, cabbage, and peas	Not reported	Pakade et al. (2013)

M. oleifera	Leaves, stems, and stalks collected at different seasons (summer, winter)	DPPH and reducing power assays	Methanolic extracts of leaves collected during winter showed higher scavenging capacity of the DPPH radical with EC_{50} = 387 µg/ml. Leaf, stem, and stalk extracts showed similar concentration-dependent reducing power	Not reported	Shih et al. (2011)
M. oleifera	Leaves	FRAP and DPPH assays	FRAP assay: 1 mg of *M. oleifera* leaf equals antioxidant values of 0.95–1.35 mmol $FeSO_4$. DPPH assay: IC_{50} values of 0.7440 mg/l	Quercetin-dirhamnosyl, quercetin-glycoside, kaempferol-glycoside, quercetin-glycoside-acetyl, kaempferol-glycoside-acetyl, quercetin-glycoside-succinoyl, kaempferol-glycoside-succinoyl	Y. Q. Wang et al. (2017)
M. oleifera	Leaves	DPPH assay and the OH-radical scavenging activity	A simulated gastrointestinal digestion was used to study the bioaccessibility of the antioxidants of *M. oleifera*. The antioxidant activity of moringa decreased after digestion	1-Caffeoylquinic acid, 5-caffeoylquinic acid, 3-caffeoylquinic acid, catechin, 6,8-di-C-glucosylapigenin, caffeic acid, rutin, quercetin-3-*O*-β-galactopyranoside, quercetin-3-*O*-β-glucoside, ferulic acid	Dou et al. (2019)

(continued)

TABLE 8.1 (Continued)
Summary of in vitro antioxidant reports of Moringa plants

Moringa source	Plant part	Methods	Description	Compounds associated	Reference
M. oleifera	Germinated seeds	ABTS and DPPH assays	An optimized process of germination significatively increases the antioxidant activity with DPPH (4869–9872 µmol TE/100 g) and ABTS (11,655–16,996 µmol TE/100 g)	Not reported	de la Mora-López et al. (2018)
M. oleifera	Leaves, seeds, and roots	DPPH, ABTS and FRAP assays	The highest DPPH radical scavenging activity and FRAP reducing power was for leaf extracts with IC_{50} of 1.02 mg/ml and 0.99 mM Fe^{+2}/g in each assay, respectively. Leaves and roots have similar antioxidant capacity by the ABTS assay with IC_{50} of 1.36 and 1.24 mg/ml, respectively	Leaves: glucomoringin, benzyl glucosinolates, 3-caffeoylquinic acid, rutin, quercetin-3-*O*-glucoside, quercetin-acetyl-glycoside, kaempferol-3-*O*-glucoside, kaempferol-acetyl-glycoside. Seeds: 3-hydroxy-4-(α-L-rhamnopyranosyloxy) benzyl glucosinolates, glucomoringin. Roots: glucomoringin, glucotropaeolin	Xu et al. (2019)
M. oleifera	Leaves	ABTS, DPPH, and -FRAP assays	Methanolic moringa extracts had higher ABTS, DPPH and FRAP antioxidant values (8.86, 1.17, and 8.01 µmol TE/ml, respectively)	Caffeic acid, gallic acid, catechin, rutin, chlorogenic acid, epigallocatechin gallate	Coz-Bolanos et al. (2018)

M. stenopela	Leaves	DPPH assay	Aqueous extract IC$_{50}$ (40 µg/ml) and methanolic extract IC$_{50}$ (36 µg/ml)	Rutin	Habtemariam and Varghese (2015)
M. peregrina	Seeds	DPPH, ABTS, superoxide anion radical scavenging assay, nitric oxide radical scavenging assay, hydrogen peroxide scavenging assay, hydroxyl radical scavenging assay	Essential oils of *M. peregrina* seeds showed antioxidant activity against all the evaluated probes in the antioxidant assays. The IC$_{50}$ values were 37.70, 34.03, 36.57, 29.15, 43.93, and 29.99 µg/ml for the DPPH, ABTS, superoxide anion, nitric oxide radical, hydrogen peroxide, and hydroxyl radical assays, respectively. *M. peregrina* essential oils are proposed as natural antioxidants in the food industry	Geijerene, linalool, caryophyllene oxide, *n*-hexadecane, and carvacrol	Senthilkumar et al. (2020)
M. stenopetala	Leaves	DPPH, FRAP, and, ferrous-ion chelating-activity assays	*M. stenopetala* extracts microencapsulated with spray and freeze-drying displayed antioxidant activity by the three methods. However, freeze-dried microencapsulates had higher antioxidant values than spry-drying microencapsulation	Not reported	Dadi et al. (2020)
M. peregrina	Aerial parts	ABTS, hydroxyl and DPPH radicals for the assays	The highest DPPH scavenging activity was shown for the leaves methanolic extracts with IC$_{50}$ values of 5.3 µg/ml. However, the ethyl acetate leaf extracts had higher antioxidant activity against the ABTS radical with IC$_{50}$ values of 49.1 µg/ml. The best antioxidants against the –OH radical were the leaf methanol extracts with IC$_{50}$ values of 76.9 µg/ml	Not reported	Al-Dabbas (2017)

(*continued*)

TABLE 8.1 (Continued)
Summary of in vitro antioxidant reports of Moringa plants

Moringa source	Plant part	Methods	Description	Compounds associated	Reference
M. stenopetala	Leaves	DPPH, hydroxyl, nitric oxide, ABTS, hydrogen peroxide radical scavenging assay, and reducing power activity assays	Methanolic extracts of leaves from *M. stenopetala* originating from the Tigray region of Ethiopia were evaluated. The authors reported dose-dependent antioxidant potential (0, 100, 200, 300, 400 and 500 µg extract/ml) which showed significant free radical scavenging activity on $NO > H_2O_2 > DPPH > NO > ABTS$, respectively	Flavonoids, terpenoids, and tannins	Hagos et al. (2018)
M. peregrina	Seeds	DPPH assay	Oil recovery from *M. peregrina* seeds was optimized by a central composition design using ultrasound-assisted extraction. The authors found that the optimal conditions are at 26.3 min for extraction time and a liquid-to-solid ratio of 17.8 ml/g. These conditions resulted in an oil extract with an improved DPPH scavenging activity of 22.08% versus 13.77% with a conventional Soxhlet method	Oleic acid and palmitic acid	Mohammadpour et al. (2019)

studies are conducted, toxicological studies are needed to test the safe use of moringa in animals and humans.

8.4.2 IN VIVO STUDIES

Moringa species have been studied *in vivo* to test the potential of their bioactive compounds by different researchers from around the world. The most studied species has been *M. oleifera*, commonly known as 'The Miracle Tree'; however, there are reports about *M. drouhardii*, *M. hildebrandtii*, *M. stenopetala*, *M. ovalifolia*, *M. peregrina*, *M. concanensis*, *M. longituba*, *M. ruspoliana*, *M. borziana*, *M. pygmaea*, *M. arborea*, and *M. rivae*.

Verma et al. (2009) evaluated the antioxidant potential of ethyl acetate extract of leaves of *M. oleifera* in rats (males of 150–200 g weight). Five groups were formed: (1) normal diet; (2) negative control, which one administrated carbon tetrachloride (CCl_4) to induce damage; (3–4), moringa extract at 50 and 100 mg/kg bw/day, respectively; and (5) vitamin E at a dose of 50 mg/kg bw/day. In all animals except group 1, damage was induced before applying the treatments; after 14 days the animals were sacrificed. The results showed that *Moringa* extract prevents damage by maintaining antioxidant enzyme levels (reduced glutathione, catalase, and SOD) in the liver and kidneys. Similarly, Olatosin et al. (2014) induced damage with CCl_4 (o induce alterations in enzymatic activity) in Wistar albino rats (males of 150–180 g weight) and were treated with 2 ml of *Moringa* oil/kg body weight for 10 days; at the end they found that the *Moringa oleifera* seed oil increase the activity of SOD (127.3 U/mg $\times$ 10^{-1}), while catalase (89.05 U/mg $\times$ 10^{-1}) remained constant in regards to control (89.09 U/mg $\times$ 10^{-1}). Likewise, malonaldehyde (MDA; a product of lipid peroxidation) was slightly diminished. These enzymes are antioxidant s, which provide important cellular defense functions. In this sense, Sharma et al. (2012) evaluated antioxidant activity of hydroethanolic extracts of *Moringa oleifera* pods (MOHE) in Swiss albino mice which were treated with 7,12-dimethylbenz[a]anthracene (DMBA) to induce damage. They administrated doses of 200 and 400 mg of MOHE extract/kg body weight daily for 14 days, compared with a synthetic antioxidant (BHT 0.5 and 1 %); their results showed that MOHE and BHT have a similar antioxidant effect, increasing the enzymatic activity of the reduced glutathione (GSH) and glutathione-S-transferase (GST) enzymes, protecting the liver against oxidative damage.

Moringa oleifera has been widely studied due to its effect against many diseases such as diabetes, a disease associated to the oxidative stress. In a study carried out in rats in which the disease was induced by the injection of Alloxan (150 mg/kg animal weight), an aqueous extract of *M. oleifera* leaves (250 mg/kg animal weight) was administered per 18 days, finding that the *Moringa* treatment maintained the levels of the enzymes SOD and catalase. In addition, it increased the GSH enzyme activity, maintaining the antioxidant system of cellular defense (Eldaim et al. 2017). Jaiswal et al. (2013) also evaluated the antioxidant effect of an aqueous extract of *M. oleifera* leaves in albino Wistar rats in which oxidative stress was induced by diabetes (streptozotocin disease inducer: STZ). They administered a dose of 200 mg/kg of aqueous extract daily for up to 21 days; later they evaluated the antioxidant activity by means of the enzymatic assays of SOD, catalase, and GST, as well as MDA. The results showed an increase in SOD, catalase, and GST activity, as well as a reduction in MDA in rats treated with *Moringa* extract. The authors mentioned that *M. oleifera* leaves are a good source of some antioxidants such as vitamins C, E, beta-carotene, phenolic acids, flavonoids, and flavonols, which gives them that antioxidant power. Likewise, Rajanandh et al. (2012) obtained the same results in SOD, catalase, and MDA when administering the same dose of *M. oleifera* leaves in Wistar rats that had induced hyperlipidemia.

In another study, an aqueous extract of *M. oleifera* leaves was administrated to goats at 200 g/head/day for 60 days; they reported an increase of GSH, catalase, and SOD, and a reduction of MDA; this antioxidant effect was associated with the higher polyphenolic content (Moyo et al. 2012). In this sense, a study in postmenopausal women (menopause is associated with oxidative stress), who were supplemented with *M. oleifera* leaves (7 g daily for 3 months), showed a decrease in lipid peroxidation and an increase in glutathione peroxidase and SOD. These enzymes decrease in menopause and there is an increase in lipid peroxidation; however, an antioxidant effect was observed by increasing the enzymes in charge of the cellular defense mechanism (Kruk et al. 2019).

Although most of the studies on *M. oleifera* have been carried out on leaves, other parts of the plant could have an antioxidant effect, as demonstrated in Adedapo et al. (2020). They carried out a recent study with methanol extracts of *M. oleifera* stem bark (50 and 100 mg/kg of animal weight), tested in adult male Wistar rats that had induced kidney damage. Their results showed an improved activity of enzymatic and non-enzymatic markers, decrease in MDA (a product of lipid peroxidation), which increases the antioxidant capacity, because ROS are reduced and inflammation markers inhibited.

Likewise, other species such as *M. peregrina* (aerial parts at 50, 150, and 500 mg/kg) have been used to fight the oxidative stress associated with cognitive functions in adult male Wistar rats. They found that a dose of 150 mg/kg had the better effect, with increased GPx antioxidant enzyme involved with the memory functions, as well as favorable changes such as an increase in glutathione (GSH) and a reduction of oxidized glutathione (GSSG), which are both biomarkers of oxidative stress (Alzoubi et al. 2017). Azim et al. (2017) evaluated the protective effect of an ethanolic extract of *M. peregrina* leaves (200 mg/kg) in albino rats (adult female) with liver injury induced by acetaminophen. The results showed a reduction of MDA, while GSH, SOD, and catalase were replenished and GPx was normalized after the treatment with the extract; because the DNA damage to the liver was reduced, the antioxidant activity has an important role in the mechanism of hepatic injury. The positive effect in this study was mainly associated with the phenolic compounds (3-OH-tyrosol, protocatechin, epicatechin, *e*-vanillic, pyrogallol, etc.) and flavonoids (rutin, naringin vitexin, and quercetin, etc.).

Moringa stenopela was evaluated in Wistar rats with induced pre-eclampsia (inducer of nitric oxide synthase, nitric oxide-nitro-L-arginine methyl ester) to reduce oxidative stress due to its contribution to atherogenicity (the increase of the lipid peroxides in the placenta during pre-eclampsia). The doses of the aqueous extract administered were 250, 500 and 1000 mg/kg/day and the response was dose-dependent with a decrease in the levels of lipid peroxides and an increase in the antioxidant activity in serum (expressed as μg of ascorbic acid/ml). These results suggest that this species could have a therapeutic action against this pathogenesis through the reduction of free radicals (Mergiaw et al. 2020).

An aqueous extract of *M. concanensis* leaves has shown potential as an antioxidant therapeutic to reduce or control oxidative damage in diabetic albio Wistar rats. The disease was induced with streptozotocin–nicotinamide, and the *Moringa* extract was administrated at 250 mg/kg for 45 days. The lipid peroxidation, evaluated by the levels of MDA, was inhibited and GSH was increased by the treatment with the extract. The authors mention that the probable mechanism of the antioxidant properties of this species is the potential to induce the GSH depletion and the reduction of free radicals in this disease (Balakrishnan et al. 2019).

Leaves of *Moringa rivae* were studied by Saleem et al. (2020b) in rats with arthritis rheumatoid (induced); they obtained methanolic and aqueous extracts of *Moringa* and were administrated at 150, 300 and 600 mg/kg/day, compared to a control (piroxicam 10 mg/kg/day), over 28 days. The results show that the *M. rivae* extracts reinstated the catalase and SOD activity in proportion to the concentration, comparable with the effect of the control treatment. A similar effect was observed in MDA, where both doses (300 and 600 mg/kg) reduced its content. Catalase and SOD are two powerful endogenous antioxidants, which act against oxidative stress. In this sense, it was found that the methanol and aqueous extracts of *M. rivae* managed to reduce it, and the authors mention that this could be a mechanism by which it inhibited the expression of genes involved in rheumatoid arthritis disease.

8.5 CONCLUSIONS

The genus *Moringa* is conformed of 13 species, *M. oleifera, M. drouhardii, M. hildebrandtii, M. stenopetala, M. ovalifolia, M. peregrina, M. concanensis, M. longituba, M. ruspoliana, M. borziana, M. pygmaea, M. arborea,* and *M. rivae.* Despite being a small group, they have been of great interest to different industries due to the great content of phytochemical compounds distributed in all parts of the plant, but mainly in seed and leaves. Such compounds are phenolic acids, flavonoids, glucosinolates and terpenoids, which have shown antioxidant properties in *in vitro* assays. However, *in vivo* assays are necessary explore more species due to there being few studies. *Moringa* species constitute an important source of phytochemical compounds with antioxidant potential.

REFERENCES

Abd Rani, N. Z., K. Husain, and E. Kumolosasi. (2018). *Moringa* genus: a review of phytochemistry and pharmacology. *Frontiers in Pharmacology* 9: 108. https://doi.org/10.3389/fphar.2018.00108.

Abo El-Fadl, S., A. Osman, A. M. Al-Zohairy, A. A. Dahab, and Z. A. Abo El Kheir. (2020). Assessment of total phenolic, flavonoid content, antioxidant potential and HPLC profile of three *Moringa* species leaf extracts. *Scientific Journal of Flowers and Ornamental Plants* 7(1): 53–70.

Adedapo, A. A., U. Etim, O. O. Falayi, B. S. Ogunpolu, T. O. Omobowale, A. A. Oyagbemi, et al. (2020). Methanol stem extract of *Moringa oleifera* mitigates glycerol-induced acute kidney damage in rats through modulation of KIM-1 and NF-kB signaling pathways. *Scientific African* 9: e00493. https://doi.org/10.1016/j.sciaf.2020.e00493.

Ahl, H. H. S., W. M. Hikal, and A. A. Mahmound. (2017). Biological actvity of *Moringa peregrina*: a review. *American Journal of Food Science and Health* 3(4): 83–87.

Al-Dabbas, M. M. (2017). Antioxidant activity of different extracts from the aerial part of *Moringa peregrina* (Forssk.) Fiori, from Jordan. *Pakistan Journal of Pharmaceutical Sciences* 30(6): 2151–2157.

Álvarez-Román, R., P. G. Silva-Flores, S. A. Galindo-Rodríguez, A. A. Huerta-Heredia, W. Vilegas, and D. Paniagua-Vega. (2020). Moisturizing and antioxidant evaluation of *Moringa oleifera* leaf extract in topical formulations by biophysical techniques. *South African Journal of Botany* 129: 404–411. https://doi.org/10.1016/j.sajb.2019. 10.011.

Alzoubi, K. H., N. Q. Rawashdeh, O. F. Khabour, T. El-Elimat, H. Albataineh, H. M. Al-Zghool, et al. (2017). Evaluation of the effect of *Moringa peregrina* extract on learning and memory: role of oxidative stress. *Journal of Molecular Neuroscience* 63(3): 355–363. https://doi.org/10.1007/s12031-017-0986-x.

Anitha, T. M., G. R. Shetty, P. C. Kumar, C. R. Pallavi, and P. E. Rajasekharan. (2018). Evaluation of bioactive compounds in leaves of *Moringa concanensis* accessions. *Journal of Pharmacognosy and Phytochemistry* SP3: 40–43.

Arora, D. S., J. G. Onsare, and H. Kaur. (2013). Bioprospecting of *Moringa* (Moringaceae): micro-biological perspective. *Journal of Pharmacognosy and Phytochemistry* 1(6): 193–215.

Asgari-Kafrani, A., M. Fazilati, and H. Nazem. (2020). Hepatoprotective and antioxidant activity of aerial parts of *Moringa oleifera* in prevention of non-alcoholic fatty liver disease in Wistar rats. *South African Journal of Botany* 129: 82–90. www.sciencedirect. com/science/article/pii/S0254629918319446.

Asghari, G., A. Palizban, and B. Bakhshaei. (2015). Quantitative analysis of the nutritional components in leaves and seeds of the Persian *Moringa peregrina* (Forssk.) Fiori. *Pharmacognosy Research* 7(3): 242.

Atawodi, S. E., J. C. Atawodi, G. A. Idakwo, B. Pfundstein, R. Haubner, G. Wurtele, et al. (2010). Evaluation of the polyphenol content and antioxidant properties of methanol extracts of the leaves, stem, and root barks of *Moringa oleifera* Lam. *Journal of Medicinal Food* 13(3): 710–716. https://doi.org/10.1089/jmf.2009.0057.

Azim, S. A. A., M. T. Abdelrahem, M. M. Said, and A. Khattab. (2017). Protective effect of *Moringa peregrina* leaves extract on acetaminophen-induced liver toxicity in albino rats. *African Journal of Traditional, Complementary Alternative Medicines* 14(2): 206–216. https://doi.org/10.21010/ajtcam.v14i2.22.

Balakrishnan, B. B., K. Krishnasamy, V. Mayakrishnan, and A. Selvaraj. (2019). *Moringa concanensis* Nimmo extracts ameliorates hyperglycemia-mediated oxidative stress and upregulates PPARγ and GLUT4 gene expression in liver and pancreas of streptozotocin-nicotinamide induced diabetic rats. *Biomedicine & Pharmacotherapy* 112: 108688. https://doi.org/10.1016/j.biopha.2019.108688.

Balamurugan, V., and V. Balakrishnan. (2013). Evaluation of phytochemical, pharmacognostical and antimicrobial activity from the bark of *Moringa concanensis* Nimmo. *International Journal of Current Microbiology and Applied Sciences* 2: 117–125.

Barakat, H., and G. A. Ghazal. (2016). Physicochemical properties of *Moringa oleifera* seeds and their edible oil cultivated at different regions in Egypt. *Food Nutrition Sciences* 7(6): 472.

Bell, L., O. O. Oloyede, S. Lignou, C. Wagstaff, and L. Methven. (2018). Taste and flavor perceptions of glucosinolates, isothiocyanates, and related compounds. *Molecular Nutrition and Food Research International* 62(18): 1700990.

Bennett, R. N., F. A. Mellon, N. Foidl, J. H. Pratt, M. S. Dupont, L. Perkins, et al. (2003). Profiling glucosinolates and phenolics in vegetative and reproductive tissues of the multi-purpose trees *Moringa oleifera* L. (horseradish tree) and *Moringa stenopetala* L. *Journal of Agricultural Food Chemistry* 51(12): 3546–3553.

Birben, E., U. M. Sahiner, C. Sackesen, S. Erzurum, and O. Kalayci. (2012). Oxidative stress and antioxidant defense. *World Allergy Organization Journal* 5(1): 9–19. https://doi.org/10.1097/WOX.0b013e3182439613.

Bohinc, T., G. S. Ban, D. Ban, and S. Trdan. (2012). Glucosinolates in plant protection strategies: a review. *Archives of Biological Sciences* 64(3): 821–828.

Buer, C. S., N. Imin, and M. A. Djordjevic. (2010). Flavonoids: new roles for old molecules. *Journal of Integrative Plant Biology* 52(1): 98–111.

Bullon, P., J. M. Morillo, M. C. Ramirez-Tortosa, J. L. Quiles, H. N. Newman, and M. Battino. (2009). Metabolic syndrome and periodontitis: is oxidative stress a common link? *Journal of Dental Research* 88 (6): 503–518. https://doi.org/10.1177/0022034509337479. https://journals.sagepub.com/doi/abs/10.1177/0022034509337479.

Cabello-Verrugio, C., C. Vilos, R. Rodrigues-Diez, and L. Estrada. (2018). Oxidative stress in disease and aging: mechanisms and therapies 2018. *Oxidative Medicine and Cellular Longevity* 2018: 2835189. https://doi.org/10.1155/2018/2835189.

Capuano, E., M. Dekker, R. Verkerk, and T. Oliviero. (2017). Food as pharma? The case of glucosinolates. *Current Pharmaceutical Design* 23(19): 2697–2721.

Cerdá, C., C. Sánchez, B. Climent, A. Vázquez, A. Iradi, F. El Amrani, et al. (2014). Oxidative stress and DNA damage in obesity-related tumorigenesis. In *Oxidative Stress and Inflammation in Non-Communicable Diseases – Molecular Mechanisms and Perspectives in therapeutics*, edited by Jordi Camps, pp. 5–17. Cham: Springer International Publishing.

Chhikara, N., A. Kaur, S. Mann, M. K. Garg, A. S. Sajad, and A. Panghal. (2021). Bioactive compounds, associated health benefits and safety considerations of Moringa oleifera L.: an updated review. *Nutrition & Food Science* 51: 255–277.

Chukwuebuka, E. (2015). *Moringa oleifera* "the mother's best friend". *International Journal of Nutrition and Food Sciences* 4(6): 624–630.

Coz-Bolanos, X., R. Campos-Vega, R. Reynoso-Camacho, M. Ramos-Gomez, G. F. Loarca-Pina, and S. H. Guzman-Maldonado. (2018). Moringa infusion (*Moringa oleifera*) rich in phenolic compounds and high antioxidant capacity attenuate nitric oxide pro-inflammatory mediator in vitro. *Industrial Crops and Products* 118: 95–101. https://doi.org/10.1016/j.indcrop.2018.03.028.

Cuellar-Núñez, M. L., G. Loarca-Pina, M. Berhow, and E. Gonzalez de Mejia. (2020). Glucosinolate-rich hydrolyzed extract from *Moringa oleifera* leaves decreased the production of TNF-α and IL-1β cytokines and induced ROS and apoptosis in human colon cancer cells. *Journal of Functional Foods* 75: 104270.

Dadi, D. W., S. A. Emire, A. D. Hagos, and J. B. Eun. (2020). Physical and functional properties, digestibility, and storage stability of spray- and freeze-dried microencapsulated bioactive products from *Moringa stenopetala* leaves extract. *Industrial Crops and Products* 156. https://doi.org/10.1016/j.indcrop.2020.112891.

de la Mora-López, G. S., J. López-Cervantes, R. Gutiérrez-Dorado, E. O. Cuevas-Rodríguez, J. Milán-Carrillo, D. I. Sánchez-Machado, et al. (2018). Effect of optimal germination conditions on antioxidant activity, phenolic content and fatty acids and amino acids profiles of *Moringa oleifera* seeds. *Revista Mexicana de Ingeniería Química* 17(2). https://doi.org/10.24275/10.24275/uam/izt/dcbi/revmexingquim/2018v17n2/Servin.

Dinesha, B. L., U. Nidoni, C. T. Ramachandra, N. Naik, and K. B. Sankalpa. (2018). Effect of extraction methods on physicochemical, nutritional, antinutritional, antioxidant and antimicrobial activity of Moringa (*Moringa oleifera* Lam.) seed kernel oil. *Journal of Applied Natural Science* 10(1): 287–295.

Dinkova-Kostova, A. T., and R. V. Kostov. (2012). Glucosinolates and isothiocyanates in health and disease. *Trends in Molecular Medicine* 18(6): 337–347.

Dou, Z., C. Chen, and X. Fu. (2019). Bioaccessibility, antioxidant activity and modulation effect on gut microbiota of bioactive compounds from *Moringa oleifera* Lam. leaves during digestion and fermentation *in vitro*. *Food & Function* 10(8): 5070–5079. https://doi.org/10.1039/C9FO00793H.

Dzieciol, M. (2020). Influence of extraction technique on yield and antioxidant activity of extracts from *Moringa oleifera* leaf. *Polish Journal of Chemical Technology* 22(4): 31–35. https://doi.org/10.2478/pjct-2020-0036.

El-Awady, M. A., M. M. Hassan, S. A. H. El-Sayed, and A. Gaber. (2016). Comparison of the antioxidant activities, phenolic and flavonoids contents of the leaves-crud extracts of *Moringa peregrine* and *Moringa oleifera*. *International Journal of Biosciences* 8(1): 55–62.

El-baset, A. (2017). Effect of foliar application of yeast extract and some of amino acids on growth and chemical composition of two drum sticks species (*Moringa oleifera* and *Moringa ovalifolia*). *Journal of Plant Production* 8(10): 953–959.

El-Hak, H. N. G., A. R. A. Moustafa, and S. R. Mansour. (2018). Toxic effect of *Moringa peregrina* seeds on histological and biochemical analyses of adult male Albino rats. *Toxicology Reports* 5: 38–45.

Eldaim, A. M. A., A. S. Abd Elrasoul, and S. A. Abd Elaziz. (2017). An aqueous extract from *Moringa oleifera* leaves ameliorates hepatotoxicity in alloxan-induced diabetic rats. *Biochemistry and Cell Biology* 95(4): 524–530. https://doi.org/10.1139/bcb-2016-0256.

Fraga, C. G., P. I. Oteiza, and M. Galleano. (2014). *In vitro* measurements and interpretation of total antioxidant capacity. *Biochimica et Biophysica Acta* 1840 (2): 931–934. https://doi.org/10.1016/j.bbagen.2013.06.030. www.sciencedirect.com/science/article/pii/S03044 16513002900.

Gandji, K., F. J. Chadare, R. Idohou, V. K. Salako, A. E. Assogbadjo, and R. L. Glèlè Kakaï. (2018). Status and utilisation of *Moringa oleifera* Lam: a review. *African Crop Science Journal* 26(1): 137–156.

Gebregiorgis, F., T. Negesse, and A. Nurfeta. (2012). Feed intake and utilization in sheep fed graded levels of dried moringa (*Moringa stenopetala*) leaf as a supplement to Rhodes grass hay. *Tropical Animal Health and Production* 44(3): 511–517. https://doi.org/10.1007/s11250-011-9927-9.

Godinez-Oviedo, A., N. Guemes-Vera, and O. A Acevedo-Sandoval. (2016). Nutritional and phytochemical composition of *Moringa oleifera* Lam and its potential use as nutraceutical plant: a review. *Pakistan Journal of Nutrition* 15(4): 397–405.

Goleniowski, M., M. Bonfill, R. Cusido, and J. Palazón. (2013). Phenolic acids. In *Natural Products: Phytochemistry, Botany and Metabolism of Alkaloids, Phenolics and Terpenes*, edited by Kishan Gopal Ramawat and Jean-Michel Mérillon (pp. 1951–1973). Berlin: Springer.

Gonzalez-Romero, J., S. Arranz-Arranz, V. Verardo, B. Garcia-Villanova, and E. J. Guerra-Hernandez. (2020). Bioactive compounds and antioxidant capacity of *Moringa* leaves grown in Spain versus 28 leaves commonly consumed in pre-packaged salads. *Processes* 8(10): 20. https://doi.org/10.3390/pr8101297.

Granato, D., F. Shahidi, R. Wrolstad, P. Kilmartin, L. D. Melton, F. J. Hidalgo, et al. (2018). Antioxidant activity, total phenolics and flavonoids contents: should we ban *in vitro* screening methods? *Food Chemistry* 264: 471–475. https://doi.org/10.1016/j.foodchem.2018.04.012. www.sciencedirect.com/science/article/pii/S03088146 18306265.

Gu, X., Y. Yang, and Z. Wang. (2020). Nutritional, phytochemical, antioxidant, α-glucosidase and α-amylase inhibitory properties of *Moringa oleifera* seeds. *South African Journal of Botany* 133: 151–160. https://doi.org/10.1016/j.sajb.2020.07.021.

Habtemariam, S., and G. K. Varghese. (2015). Extractability of rutin in herbal tea preparations of *Moringa stenopetala* leaves. *Beverage* 1(3): 169–182. https://doi.org/10.3390/beverages1030169. www.mdpi.com/2306-5710/1/3/169.

Hagos, Z., M. Z. Teka, M. Y. Brhane, V. K. Gopalakrishnan, and K. K. Chaithanya. (2018). *In vitro* antioxidant activities of the methanolic and aqueous extracts of *Moringa stenopetala* leaves. *Drug Invention Today* 10(5): 642–650. https://search.ebscohost.com/login.aspx?direct=true&db=asn&AN=130155856&site=ehost-live.

Halliwell, B., and J. M. C. Gutteridge. (2015). *Free Radicals in Biology and Medicine.* New York: Oxford University Press.

Hausiku, M. K., E. G. Kwembeya, P. M. Chimwamurombe, and A. Mbangu. (2020). Assessment of species boundaries of the *Moringa ovalifolia* in Namibia using nuclear its DNA sequence data. *South African Journal of Botany* 131: 335–341.

Heleno, S. A., A. Martins, M. J. R. P. Queiroz, and I. C. F. R. Ferreira. (2015). Bioactivity of phenolic acids: metabolites versus parent compounds: a review. *Food Chemistry* 173: 501–513.

Jaiswal, D., P. K. Rai, S. Mehta, S. Chatterji, S. Shukla, D. K. Rai, et al. (2013). Role of *Moringa oleifera* in regulation of diabetes-induced oxidative stress. *Asian Pacific Journal of Tropical Medicine* 6(6): 426–432. https://doi.org/10.1016/S1995-7645(13)60068-1.

Juhaimi, F. A. L., K. Ghafoor, I. A. Mohamed Ahmed, E. E. Babiker, and M. M. Özcan. (2017). Comparative study of mineral and oxidative status of *Sonchus oleraceus*, *Moringa oleifera* and *Moringa peregrina* leaves. *Journal of Food Measurement Characterization* 11(4): 1745–1751.

Khalid, M., M. Bilal, H. M. N. Iqbal, and D. Huang. (2019). Biosynthesis and biomedical perspectives of carotenoids with special reference to human health-related applications. *Biocatalysis Agricultural Biotechnology* 17: 399–407.

Kruk, J., H. Y. Aboul-Enein, A. Kładna, and J. E. Bowser. (2019). Oxidative stress in biological systems and its relation with pathophysiological functions: the effect of physical activity on cellular redox homeostasis. *Free Radical Research* 53(5): 497–521. https://doi.org/10.1080/10715762.2019.1612059.

Kumar, G., A. A. Giri, R. Arya, R. Tyagi, S. Mishra, and A. K. Mishra. (2019). Multifaceted applications of different parts of *Moringa* species: review of present status and future potentials. *International Journal of Chemistry Studies* 7(2): 835–842.

Kumar, N., and N. Goel. (2019). Phenolic acids: natural versatile molecules with promising therapeutic applications. *Biotechnology Reports* 24: e00370.

Kumar, S., and A. K. Pandey. (2013). Chemistry and biological activities of flavonoids: an overview. *The Scientific World Journal* 2013: 1–16.

Landazuri, A. C., A. Gualle, V. Castaneda, E. Morales, A. Caicedo, and L. M. Orejuela-Escobar. (2020). *Moringa oleifera* Lam. leaf powder antioxidant activity and cytotoxicity in human primary fibroblasts. *Natural Product Research*: 6. https://doi.org/10.1080/14786419.2020.1837804.

Leone, A., A. Spada, A. Battezzati, A. Schiraldi, J. Aristil, and S. Bertoli. (2015). Cultivation, genetic, ethnopharmacology, phytochemistry and pharmacology of *Moringa oleifera* leaves: an overview. *International Journal of Molecular Sciences* 16(6).

Liu, R. H. (2013). Dietary bioactive compounds and their health implications. *Journal of Food Science* 78(s1): A18–A25.

Luqman, S., S. Srivastava, R. Kumar, A. KumarMaurya, and D. Chanda. (2012). Experimental assessment of *Moringa oleifera* leaf and fruit for its antistress, antioxidant, and scavenging potential using *in vitro* and *in vivo* assays. *Evidence-Based Complementary and Alternative Medicine* 2012: 519084. https://doi.org/10.1155/2012/519084.

Ma, Z. F., J. Ahmad, H. Zhang, I. Khan, and S. Muhammad. (2020). Evaluation of phytochemical and medicinal properties of moringa (*Moringa oleifera*) as a potential functional food. *South African Journal of Botany* 129: 40–46.

Makita, C., L. Chimuka, P. Steenkamp, E. Cukrowska, and E. Madala. (2016). Comparative analyses of flavonoid content in *Moringa oleifera* and *Moringa ovalifolia* with the aid of UHPLC-qTOF-MS fingerprinting. *South African Journal of Botany* 105: 116–122. https://doi.org/10.1016/j.sajb.2015.12.007.

Makita, C., L. Chimuka, E. Cukrowska, P. A. Steenkamp, M. Kandawa-Schutz, A. R. Ndhlala, et al. (2017). UPLC-qTOF-MS profiling of pharmacologically important chlorogenic acids and associated glycosides in *Moringa ovalifolia* leaf extracts. *South African Journal of Botany* 108: 193–199. https://doi.org/10.1016/j.sajb.2016.10.016.

Maldini, M., S. A. Maksoud, F. Natella, P. Montoro, G. L. Petretto, M. Foddai, et al. (2014). *Moringa oleifera*: study of phenolics and glucosinolates by mass spectrometry. *Journal of Mass Spectrometry* 49(9): 900–910.

Martínez-Navarrete, N., M. del Mar Camacho Vidal, and J. J. Martínez Lahuerta. (2008). Los compuestos bioactivos de las frutas y sus efectos en la salud. *Actividad Dietética* 12(2): 64–68.

Mekonnen, Y., and B. Dräger. (2003). Glucosinolates in *Moringa stenopetala*. *Planta Medica* 69(4): 380–382.

Mergiaw, K., Y. A. Mengesha, T. Tolessa, E. Makonnen, S. Genet, A. Abebe, et al. (2020). Effects of *Thymus schimperi* and *Moringa stenopetala* leaf extracts on lipid peroxidation and total antioxidant status in pre-eclampsia rat models. *Journal of Complementary Alternative Medical Research* 10(2): 1–7. https://doi.org/10.9734/JOCAMR/2020/v10i230157.

Mohammadpour, H., S. M. Sadrameli, F. Eslami, and A. Asoodeh. (2019). Optimization of ultrasound-assisted extraction of *Moringa peregrina* oil with response surface methodology and comparison with Soxhlet method. *Industrial Crops and Products* 131: 106–116. https://doi.org/10.1016/j.indcrop.2019.01.030.

Moyo, B., S. Oyedemi, P. J. Masika, and V. Muchenje. (2012). Polyphenolic content and antioxidant properties of *Moringa oleifera* leaf extracts and enzymatic activity of liver from goats supplemented with *Moringa oleifera* leaves/sunflower seed cake. *Meat Science* 91(4): 441–447. https://doi.org/10.1016/j.meatsci.2012.02.029.

Ndhlala, A. R., R. Mulaudzi, B. Ncube, H. A. Abdelgadir, C. P. du Plooy, and J. Van Staden. (2014). Antioxidant, antimicrobial and phytochemical variations in thirteen *Moringa oleifera* Lam. cultivars. *Molecules* 19(7): 10480–10494. https://doi.org/10.3390/molecules190710480.

Oguntibeju, O. O., G. Y. Aboua, and E. I. Omodanisi. (2020). Effects of *Moringa oleifera* on oxidative stress, apoptotic and inflammatory biomarkers in streptozotocin-induced diabetic animal model. *South African Journal of Botany* 129: 354–365.

Olatosin, T. M., D. S. Akinduko, and C. Z. Uche. (2014). Antioxidant capacity of *Moringa oleifera* seed oil against CCl_4 induced hepatocellular lipid peroxidation in Wistar albino rats. *European Journal of Experimental Biology* 4(1): 514–518.

Olson, M. E. (2019). Introduction to the Moringa family: Origin, distribution and biodiversity. In *The Miracle Tree: Moringa oleifera*, edited by M. Palada, A. W. Ebert and R. C. Joshi. US: Xlibris Corporation.

Padayachee, B., and H. Baijnath. (2012). An overview of the medicinal importance of Moringaceae. *Journal of Medicinal Plants Research* 6(48): 5831–5839. https://doi.org/ 10.5897/JMPR12.1187.

Pakade, V., E. Cukrowska, and L. Chimuka. (2013). Comparison of antioxidant activity of *Moringa oleifera* and selected vegetables in South Africa. *South African Journal of Science* 109(3–4): 17–21. https://doi.org/10.1590/sajs.2013/1154. <Go to ISI>:// WOS:000323295300006.

Palada, M. C. (2019). Introduction: *Moringa* – a miracle tree crop. In *The Miracle Tree: Moringa oleifera*, edited by M. Palada, A. W. Ebert and R. C. Joshi. US: Xlibris Corporation.

Panche, A. N., A. D. Diwan, and S. R. Chandra. (2016). Flavonoids: an overview. *Journal of Ntritional Science* 5: e47.

Rajanandh, M. G., M. N. Satishkumar, K. Elango, and B. Suresh. (2012). *Moringa oleifera* Lam. A herbal medicine for hyperlipidemia: a pre–clinical report. *Asian Pacific Journal of Tropical Disease* 2: S790–S795. https://doi.org/10.1016/S2222-1808(12)60266-7.

Rashmi, H. B., and P. S. Negi. (2020). Phenolic acids from vegetables: a review on processing stability and health benefits. *Food Research International* 136: 109298.

Razis, A. F. A., M. Din Ibrahim, and S. B. Kntayya. (2014). Health benefits of *Moringa oleifera*. *Asian Pacific Journal of Cancer Prevention* 15 (20): 8571–8576.

Ribeiro, D., M. Freitas, A. M. S. Silva, F. Carvalho, and E. Fernandes. (2018). Antioxidant and pro-oxidant activities of carotenoids and their oxidation products. *Food Chemical Toxicology* 120: 681–699.

Robiansyah, I., A. S. Hajar, M. A. Al-kordy, and A. Ramadan. (2014). Current status of economically important plant *Moringa peregrina* (Forrsk.) Fiori in Saudi Arabia: a review. *International Journal of Theoretical and Applied Sciences* 6(1): 79.

Roohbakhsh, A., G. Karimi, and M. Iranshahi. (2017). Carotenoids in the treatment of diabetes mellitus and its complications: a mechanistic review. *Biomedicine Pharmacotherapy* 91: 31–42.

Rowles III, J. L., and J. W. Erdman Jr. (2020). Carotenoids and their role in cancer prevention. *Biochimica et Biophysica Acta – Molecular Cell Biology of Lipids* 1865: 158613.

Sahay, S., U. Yadav, and S. Srinivasamurthy. (2017). Potential of *Moringa oleifera* as a functional food ingredient: a review. *International Journal of Food Science and Nutrition* 2(5): 31–37.

Said-Al Ahl, H. A. H., W. M. Hikal, and A. A. Mahmoud. (2017). Biological activity of *Moringa peregrina*: a review. *American Journal of Food Science and Health* 3(4): 83–87.

Saini, R. K., N. P. Shetty, and P. Giridhar. (2014). Carotenoid content in vegetative and reproductive parts of commercially grown *Moringa oleifera* Lam. cultivars from India by LC–APCI–MS. *European Food Research Technology* 238(6): 971–978.

Saini, R. K., N. P. Shetty, M. Prakash, and P. Giridhar. (2014). Effect of dehydration methods on retention of carotenoids, tocopherols, ascorbic acid and antioxidant activity in *Moringa oleifera* leaves and preparation of a RTE product. *Journal of Food Science Technology* 51(9): 2176–2182.

Salaheldeen, M., M. K. Aroua, A. A. Mariod, S. F. Cheng, and M. A. Abdelrahman. (2014). An evaluation of *Moringa peregrina* seeds as a source for bio-fuel. *Industrial Crops and Products* 61: 49–61.

Saleem, A., M. Saleem, and M. F. Akhtar. (2020). Antioxidant, anti-inflammatory and antiarthritic potential of *Moringa oleifera* Lam: an ethnomedicinal plant of Moringaceae family. *South African Journal of Botany* 128: 246–256.

Saleem, A., M. Saleem, M. F. Akhtar, M. Shahzad, and S. Jahan. (2020). *Moringa rivae* leaf extracts attenuate Complete Freund's adjuvant-induced arthritis in Wistar rats via modulation of inflammatory and oxidative stress biomarkers. *Inflammopharmacology* 28(1): 139–151. https://doi.org/10.1007/s10787-019-00596-3.

Samanta, A., G. Das, and S. K. Das. (2011). Roles of flavonoids in plants. *International Journal of Pharmaceutical Science and Technology* 100(6): 12–35.

Sánchez-Machado, D. I., J. López-Cervantes, J. A. Núñez-Gastélum, G. S. de la Mora-López, J. López-Hernández, and P. Paseiro-Losada. (2015). Effect of the refining process on *Moringa oleifera* seed oil quality. *Food Chemistry* 187: 53–57.

Sánchez-Pujante, P. J., M. Borja-Martínez, M. Á. Pedreño, and L. Almagro. (2017). Biosynthesis and bioactivity of glucosinolates and their production in plant in vitro cultures. *Planta* 246(1): 19–32.

Santhi, K., and R. Sengottuvel. (2016). Qualitative and quantitative phytochemical analysis of *Moringa concanensis* Nimmo. *International Journal of Current Microbiology and Applied Sciences* 5(1): 633–640.

Sarwar, M. (2017). Biodiversity and conservation of *Moringa* species. *Journal of Biodiversity and Conservation* 1(1): 5–7.

Senthilkumar, A., A. Thangamani, K. Karthishwaran, and A. J. Cheruth. (2020). Essential oil from the seeds of *Moringa peregrina*: chemical composition and antioxidant potential. *South African Journal of Botany* 129: 100–105. https://doi.org/10.1016/j.sajb.2019.01.030.

Senthilkumar, A., N. Karuvantevida, L. Rastrelli, S. S. Kurup, and A. J. Cheruth. (2018). Traditional uses, pharmacological efficacy, and phytochemistry of *Moringa peregrina* (Forssk.) Fiori. – a review. *Frontiers in Pharmacology* 9: 465.

Shahzad, U., M. A. Khan, M. J. Jaskani, I. A. Khan, and S. S. Korban. (2013). Genetic diversity and population structure of *Moringa oleifera*. *Conservation Genetics* 14(6): 1161–1172.

Sharma, V., R. Paliwal, P. Janmeda, and S. Sharma. (2012). Chemopreventive efficacy of *Moringa oleifera* pods against 7,12-dimethylbenz[*a*]anthracene induced hepatic carcinogenesis in mice. *Asian Pacific Journal of Cancer Prevention* 13(6): 2563–2569. https://doi.org/10.7314/APJCP.2012.13.6.2563.

Shih, M. C., C. M. Chang, S. M. Kang, and M. L. Tsai. (2011). Effect of different parts (leaf, stem and stalk) and seasons (summer and winter) on the chemical compositions and antioxidant activity of *Moringa oleifera*. *International Journal of Molecular Sciences* 12(9): 6077–6088. https://doi.org/10.3390/ijms12096077.

Singh, B., and R. A. Sharma. (2020). *Moringa* species. In *Secondary Metabolites of Medicinal Plants*, edited by B. Singh and R. A. Sharma (pp. 699–711). Weinheim: Wiley.

Singh, R. S. G., P. S. Negi, and C. Radha. (2013). Phenolic composition, antioxidant and antimicrobial activities of free and bound phenolic extracts of *Moringa oleifera* seed flour. *Journal of Functional Foods* 5(4): 1883–1891. https://doi.org/10.1016/j.jff.2013.09.009.

Teixeira, E. M. B., M. R. Barbieri Carvalho, V. A. Neves, M. A. Silva, and L. Arantes-Pereira. (2014). Chemical characteristics and fractionation of proteins from *Moringa oleifera* Lam. leaves. *Food Chemistry* 147: 51–54.

Velázquez-Zavala, M., I. E. Peón-Escalante, R. Zepeda-Bautista, and M. A. Jiménez-Arellanes. (2016). Moringa (*Moringa oleifera* Lam.): potential uses in agriculture, industry and medicine. *Revista Chapingo. Serie horticultura* 22(2): 95–116.

Verma, A. R., M. Vijayakumar, C. S. Mathela, and C. V. Rao. (2009). *In vitro* and *in vivo* antioxidant properties of different fractions of *Moringa oleifera* leaves. *Food and Chemical Toxicology* 47(9): 2196–2201. https://doi.org/10.1016/j.fct.2009.06.005. www.sciencedirect.com/science/article/pii/S0278691509002774.

Veskoukis, A. S., A. M. Tsatsakis, and D. Kouretas. (2012). Dietary oxidative stress and antioxidant defense with an emphasis on plant extract administration. *Cell Stress and Chaperones* 17(1): 11–21. https://doi.org/10.1007/s12192-011-0293-3.

von Lintig, J., J. Moon, J. Lee, and S. Ramkumar. (2019). Carotenoid metabolism at the intestinal barrier. *Biochimica et Biophysica Acta – Molecular Cell Biology of Lipids* 1865: 158580.

Wang, F., S. Long, J. Zhang, J. Yu, Y. Xiong, W. i Zhou, et al. (2020). Antioxidant activities and anti-proliferative effects of *Moringa oleifera* L. extracts with head and neck cancer. *Food Bioscience* 37: 100691.

Wang, Y. Q., Y. J. Gao, H. Ding, S. J. Liu, X. Han, J. Z. Gui, et al. (2017). Subcritical ethanol extraction of flavonoids from *Moringa oleifera* leaf and evaluation of antioxidant activity. *Food Chemistry* 218: 152–158. https://doi.org/10.1016/j.foodchem.2016.09.058.

Weaver, C. M. (2014). Bioactive foods and ingredients for health. *Advances in Nutrition* 5(3): 306S–311S.

Xiao, Z.-P., Z.-Y. Peng, M.-J. Peng, W.-B. Yan, Y.-Z. Ouyang, and H.-L. Zhu. (2011). Flavonoids health benefits and their molecular mechanism. *Mini Reviews in Medicinal Chemistry* 11(2): 169–177.

Xu, Y. B., G. L. Chen, and M. Q. Guo. (2019). Antioxidant and anti-inflammatory activities of the crude extracts of *Moringa oleifera* from Kenya and their correlations with flavonoids. *Antioxidants* 8(8): 12. https://doi.org/10.3390/antiox8080296.

Yeum, K.-J., R. M. Rusell, and G. Aldini. (2010). Antioxidant activity and oxidative stress: an overview. In *Biomarkers for Antioxidant Defense and Oxidative Damage: Principles and Practical Applications*, edited by Giancarlo Aldini, Kyung-Jin Yeum, Estuo Niki and Robert M. Rusell (pp. 3–19). Ames: Wiley-Blackwell.

9 Anti-Inflammatory Properties of *Moringa oleifera*

Rosmarbel Morales-Nava
Tecnologico de Monterrey, Av. Eugenio Garza Sada 2501 Sur, Col. Tecnologico, Monterrey, N.L., México CP 64849.

Janet Alejandra Gutiérrez-Uribe
Tecnologico de Monterrey, Av. Eugenio Garza Sada 2501 Sur, Col. Tecnologico, Monterrey, N.L., México CP 64849.
Corresponding author e-mail: jagu@tec.mx

CONTENTS

9.1 INTRODUCTION: BACKGROUND AND DRIVING FORCES

Moringa oleifera is a native plant from Himalayas and India belonging to the family Moringaceae. It has been called "The Miracle Tree" or "The Tree of the Life" because all parts of the plant are useful for humans. It is commonly known in English as drumstick tree due to its long and slender seed pod appearance, horseradish tree from the roots' spicy taste, and ben oil or benzoyl tree because of the oil extracted from the seeds. The genus *Moringa* is made up of 13 species that are widely cultivated in Asia and Africa. However, its cultivation has also spread in the tropical and sub-tropical areas in Central America and the Caribbean. *Moringa oleifera* is also naturalized in different climates, adapted even to harsh and dry soils.

DOI: 10.1201/9781003108863-9

FIGURE 9.1   Aerial parts of *Moringa oleifera*. (a) Leaves, (b) bark/wood, (c) flowers, (d) seeds pods.

Moringa oleifera is part of a genus of important medicinal plants that have been used for a wide variety of ailments in traditional medicine, including colds, sore throats, anemia, stress, depression, diuretic, and antispasmodic, besides its use as an antimicrobial and antibacterial. *Moringa* has also been used to treat diseases such as malnutrition, diabetes, hypertension, arthritis, and kidney stone disorders, and more recently, its anticancer properties have been explored (Abd Rani et al. 2018; Srivastava et al. 2020; Mishra et al. 2011).

Almost every part of this plant (leaves, bark, roots, flowers, and seeds; Figure 9.1) has been found to exhibit biological activities. Among these, its activity as blood glucose regulator, antioxidant, pain relief, anti-inflammatory, and anticancer stands out and has an important role in regulating the urinary tract and as an adjuvant in breastfeeding (Meireles et al., 2020).

In recent years, research into the anti-inflammatory properties of *M. oleifera* has taken on special relevance, where the incorporation of *in vitro* and *in vivo* studies has allowed us to understand the anti-inflammatory activity of *Moringa* better and correlate this to the chemical compounds responsible. Inflammation is a primary defense mechanism that the body uses as protection against harmful stimuli such as burns, allergens, infections, toxins, etc. Chronic inflammation can act as a predisposing factor for the development of other chronic diseases. Due to the impact of inflammatory processes on health, the pharmaceutical industry has developed drugs to control it; however, it has been reported that a large number of these drugs have severe collateral effects. The use of medicinal plants such as *M. oleifera* is a feasible alternative for study of the control of inflammatory processes. In this chapter, we present some of the most relevant findings from these studies. Kou et al. (2018) present a review of a wide variety of studies that demonstrate the efficacy of *M. oleifera* extracts to treat inflammatory processes and their related biomolecular markers.

Among the main compounds found to contribute to the anti-inflammatory properties of *Moringa* are tannins, polyphenols, alkaloids, carotenoids, flavonoids, moringine, moringinine, β-sitosterol, β-sitostenone, and 9-octadecenoic acid (Leone et al. 2015). The occurrence of these chemical compounds is different in leaves, seeds, bark, roots, and flowers, and can vary depending on cultivation conditions, soil nutrients, harvest season, and other factors (Venkataswera et al. 1999; Sharma et al. 2011; Saini et al. 2016; Kumar 2017; Bhattacharya et al. 2018; Srivastava et al. 2020). This chapter presents a general review of the *in vitro* and *in vivo* studies done in different parts of the *Moringa oleifera* plant, focusing on the anti-inflammatory effects of the extracts.

9.2 *IN VITRO* STUDIES

The *in vitro* tests of *Moringa oleifera* are diversified in the study of both the crude extracts and compounds isolated from the extracts. *In vitro* studies include the protein denaturation inhibition assay. As shown in Figure 9.2, the protein denaturation method is used to evaluate the capacity of chemical compounds to inhibit the denaturation of a protein (normally egg albumin) under stress conditions such as heat or oxidative stress. An ulterior spectrometric analysis of absorbance determines if the chemical compound or extract used with the protein could inhibit the denaturation.

The anti-inflammatory activity of *Moringa* could also be tested by measuring the effect of the extract against the denaturation of protein compared with standard anti-inflammatory as shown by the study of Shallangwa et al. (2016). In this study, the denaturation protective effect of the ethanol extract of leaves of *M. oleifera* was tested and correlated with phenolic compounds and flavonoids contents. Here, the extract was incubated with egg albumin, and the absorbance was determined to compare its value with the presented by the experiment made with the standard anti-inflammatory drug ibuprofen. The results showed that the denaturation protein has a concentration-dependent behavior having an $EC_{50} = 215.9 \pm 29.8$ µg/ml to ethanolic extract, compared with ibuprofen at $EC_{50} = 1599 \pm 337$ µg/ml. This is correlated with the antioxidant activity due to its total flavonoid contents of 5.82 ± 1.38 µg. This indicates that the ethanolic extract of *Moringa* has an important *in vitro* anti-inflammatory activity against the denaturation of protein compared with the standard anti-inflammatory.

9.2.1 *MORINGA OLEIFERA* LEAVES

Moringa leaves contain phytochemicals such as acetyl-glucomoringin, caffeoylquinic acid, feruloylquinic acid, and coumarylquinic acid, which have been related to the

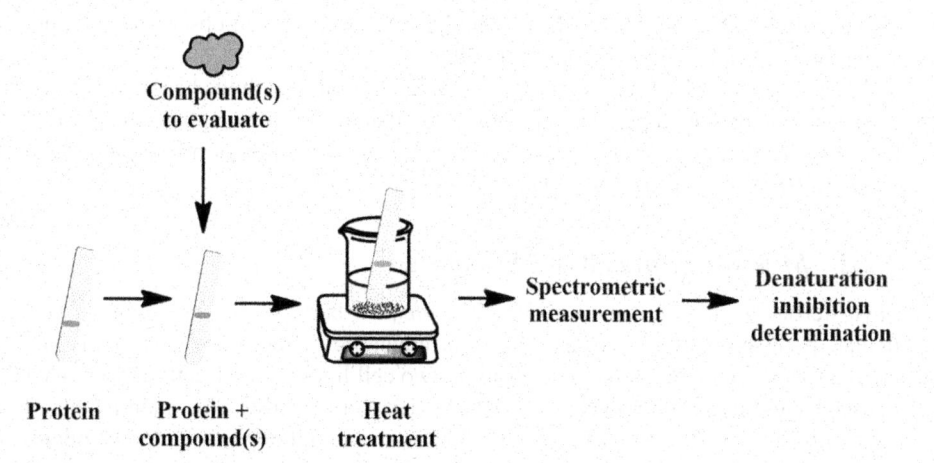

Compound(s)
to evaluate

Protein   Protein +   Heat   Spectrometric   Denaturation
           compound(s) treatment measurement   inhibition
                                                   determination

FIGURE 9.2 Protein denaturation inhibition assay used to test the bioactivity of *Moringa* extracts.

antioxidant, antibacterial, and anti-inflammatory activities of the acetonic extract of this aerial part (Sahakitpichan et al. 2011; Yan et al. 2020). *Moringa* leaves also contain flavonoids, saponins, tannins, terpenoids, proanthocyanidins, and cardiac glycosides, alkaloids, amino acids, sterols, carbohydrates, iron, calcium, phosphorus, vitamins A and B, α-tocopherol, riboflavin, nicotinic acid, folic acid, among others (Srivastava et al. 2020).

To test the anti-inflammatory activity, it is common to use high-polarity extracts of *M. oleifera* leaves as ethanolic, methanolic, and aqueous extracts. The high-polarity extracts have a wider variety of molecules from low to high polarity resulting in a rich assortment of chemical compounds. When tested in RAW 264.7 cells, an acetonic extract of *Moringa* leaves inhibited the expression of inducible NO synthase (iNOS) mRNA levels when the cells were stimulated with lipopolysaccharide (Yan et al. 2020). Interestingly, *Moringa* leaves have been roasted to improve the extraction of phytochemicals. Fombang et al. (2020) used roasted leaves to maximize the extraction yield and phenolic content simultaneously. Using 65%–75% ethanol at 55–65°C, the extracts obtained after 40 min of mixing were shown to contain 45 mg/g of rutin as the major phenolic compound along with quercetin, isoquercetin, and caffeic acid (Fombang et al. 2020). Rutin is a quercetin glycoside that has been correlated with anti-inflammatory activity as a pure compound or found in extracts from different plants (Table 9.1). Other compounds have been found using mass spectrometry, other phytochemicals such as acetyl-glucomoringin, caffeoylquinic acid, feruloylquinic acid, and coumarylquinic acid, which are in high concentrations in *Moringa* leaf acetonic extract (Yan et al. 2020). Kaempferol is another flavonol found in *Moringa* leaves at higher concentrations than quercetin and its glycosides (Saleem et al. 2020). Kaempferol and its glycosides also reduced colon inflammation in mice (Zhang et al. 2020). As can be seen in Table 9.1, kaempferol as a pure compound has been proved to downregulate NF-κB, p-NF-κB, and p-GSK-3β, suppress the levels of TNF-α, iNOS, IL-12, and IL-6 (Wang et al. 2020a; Wang et al. 2020b; Yang et al. 2020; Yao et al. 2020), and decrease the levels of IL-5, IL-13, GM-CSF, and eosinophil count (Molitorisova et al. 2021). Similar effects have been observed with pure quercetin (Beken et al. 2020; Yuan et al. 2020).

Besides flavonoids, glucosinolates have been correlated with the beneficial effects of *Moringa* leaves. Compared to aqueous or methanolic extracts, a glucosinolate-rich hydrolyzed extract of *Moringa* leaves showed lower production of pro-inflammatory cytokines IL-1β and TNF-α (Cuellar-Núñez et al. 2020).

9.2.2 *MORINGA OLEIFERA* BARK/WOOD/STEM

Phytosterols are mainly found in *Moringa oleifera* bark, wood, and stem; in particular, β-sitosterol was recovered from an ethanolic extract of woody stems and proven to have anti-inflammatory activity in two cell lines (HaCaT cells and J774A.1 macrophages) in a dose range of 7.5–30 μM (Liao et al. 2018). The main results of this study show that in HaCaT cells, the β-sitosterol could downregulate the inflammatory cytokines IL-6 and IL-8, reduce the production of reactive oxygen species (ROS), and induce the generation of an anti-inflammatory protein, the HO-1, besides

TABLE 9.1
Overview of chemical compounds with anti-inflammatory properties present in *Moringa oleifera*

Common chemical constituents	Anti-inflammatory activity	Tested with	Reference
β-sitosterol	NO inhibition,	*Moringa* extract	Liao et al. 2018
	Downregulated IL-6 and IL-8	Pure compound	Kurano et al. 2018
	Reduce serum levels of IL-6 and TNF-α	Pure compound	Cheng et al. 2020
	Increased the expression of arginase-1, IL-10, CD163 and CD206	Pure compound Pure compound	Liu et al. 2019 Koc et al. 2021
	Downregulated the levels of TNF-α and IL-6 and upregulated the levels of endothelial NOS	Pure compound	Ding et al. 2019 Zhou et al. 2020
	Decreased the levels of TNF-α, IL-6, and IL-1β		
	Supressed NF-κB and p38 mitogen-activated protein kinase (MAPK) signaling		
Kaempferol	Inhibited nitric oxide (NO) production and inducible NO synthase (iNOS) mRNA levels	*Moringa* leaf acetonic extract	Yan et al. 2020
	Inhibition of paw inflammation	Semipurified leaf methanolic extract	Saleem et al. 2020
	Inhibited the secretion of IL-6 and TNF-α.	Leaf ethanolic extract	Zhang et al. 2020
	Downregulation of NF-κB, p-NF-κB, and p-GSK-3β	Pure compound	Wang et al. 2020a
	Suppressed the levels of TNF-α, iNOS, IL-12	Pure compound	Wang et al. 2020b

β-sitosterol

Kaempferol

(*continued*)

TABLE 9.1 (Continued)
Overview of chemical compounds with anti-inflammatory properties present in Moringa oleifera

Common chemical constituents	Anti-inflammatory activity	Tested with	Reference
	Suppressed the levels of TNF-α and IL-6, and the activation of NF-κB	Pure compound	Yao et al. 2020
	Inhibit release of inflammatory factors including IL-6 and TNF-α	Pure compound	Yang et al. 2020
	Decreased the levels of IL-5, IL-13, GM-CSF and eosinophil count	Pure compound	Molitorisova et al. 2021
	Reduced the expression of IL-1β, IL-6 and IL-8	Pure compound	Beken et al. 2020
	Inhibited neutrophil infiltration and reduced the plasma levels of inflammatory cytokines.	Pure compound	Yuan et al. 2020

Quercetin

Quercetin-3-O-glucoside

(continued)

TABLE 9.1 (Continued)

Overview of chemical compounds with anti-inflammatory properties present in Moringa oleifera

Common chemical constituents	Anti-inflammatory activity	Tested with	Reference
R_1-R_4 = H, OH, CH$_3$ **Glucomoringin isothiocyanate derivatives**	Decreased the production of TNF-α and IL-1β	Leaf methanolic extract enriched in hydrolyzed glucosinolates	Cuellar-Nuñez et al. 2020
α-tocopherol	Lowest edema in rats induced with carrageenan	Dichlorometane fraction of an ethanolic extract of *Moringa* leaves	Mahdi et al. 2017

the inhibition of other pro-inflammatory molecular factors. As can be observed in Table 9.1, isolated β-sitosterol reduces serum levels of IL-6 and TNF-α, increases the expression of arginase-1, IL-10, CD163, and CD206, and suppresses NF-κB and p38 mitogen-activated protein kinase (MAPK) signaling (Ding et al. 2018; Kurano et al. 2018; Liu et al. 2019; Zhou et al. 2020; Koc et al. 2021).

Sitosterol is found in other plants and it has been proved in animal models to improve immune function and inflammation, particularly at a dose higher than 60 mg/kg it decreased pro-inflammatory cytokine level in broilers' intestine (Cheng et al. 2020).

9.2.3 *MORINGA OLEIFERA* ROOTS

A recent evaluation of extracts of *M. oleifera* roots has shown the importance of the novel polysaccharide MRP-1 as an anti-inflammatory agent (Cui et al. 2019). This novel compound was isolated from hot aqueous media and ethanolic precipitation. The monosaccharide composition obtained by gas chromatography–mass spectrometry (GC-MS) showed that MRP-1 mainly consisted of rhamnose, arabinose, fructose, xylose, mannose, and galactose in a molar ratio of 1.5:2.0:3.1:6.0:5.3:1.1. In RAW264.7 macrophage cells, MRP-1 prevented the increase of NO and TNF-α production induced by lipopolysaccharide (LPS). Moreover, the mRNA expression level of iNOS induced by LPS was significantly decreased ($p < 0.05$) while showing no obvious effect on the COX-2 mRNA expression. Although roots may accumulate toxic contaminants from soil or water, some alternatives have been developed to use this part of the plant to produce polysaccharide extracts enriched in selenium and lower cadmium contents (Fu et al. 2020).

9.2.4 *MORINGA OLEIFERA* FLOWERS AND PODS

An evaluation of *M. oleifera* flowers was presented by Alhakmani et al. (2013). The ethanolic flowers extract was obtained by cold maceration, total phenolics were estimated with Folin–Ciocalteu reagent/spectrometric UV-VIS analysis, and the anti-inflammatory activity was determined by protein denaturation method and compared to diclofenac sodium effect. Among the phytochemicals found in the ethanolic extract were flavonoids, tannins, cardiac glycosides, and alkaloids. The total phenolics content was 19.31 mg/g quantified as gallic acid equivalents. The extract's anti-inflammatory effect was determined using egg albumin and heat treatment to measure the absorbance at 660 nm and correlate it with the denaturation inhibition percentage caused by the extract (Figure 9.2). The *M. oleifera* flowers extract (100–500 μg/ml) gave a significant dose-dependent inhibition to albumin denaturation. The extract's inhibition at 200 μg/ml was comparable with the effect of diclofenac sodium at 100 μg/ml (88.10% and 84.95%, respectively). This anti-inflammatory effect correlated to the high phenolic content and hydroethanolic extract of flowers has also proven to reduce inflammatory markers in animal models with induced hepatotoxicity when tested at 200 and 400 mg/kg (Sharifudin et al. 2013).

Another aerial part that has been traditionally consumed and proven to have significant inflammation effects is the pods. Combined aqueous and ethanolic extracts were tested in RAW 264.7 macrophage cells, demonstrating the expression and protein levels of interleukine-6, tumor necrosis factor-alpha, iNOS, and cyclooxygenase-2 that was mediated partly by inhibiting phosphorylation of inhibitor kappa B protein and mitogen-activated protein kinases (Muangnoi et al. 2012).

9.3 *IN VIVO* STUDIES

The carrageenan-induced edema method is a very useful technique to measure the anti-inflammatory activity in *in vivo* studies in an animal model with mice or rats. Normally, the animals are administrated with extracts before the chemical induction of edema; carrageenin is commonly used to induce it. The decrease of the induced edema is compared with the effect produced by anti-inflammatory drugs or extracts, as shown in Figure 9.3. Other animal models use the *in situ* expositions of inflammatory agents like chemical compounds or allergens to produce localized inflammation and probe the effect of different extracts. In some studies, the characterization of the chemical components present in the extracts complements the results and provides us with information to relate the chemical structure with the biological activity presented. More recently, molecular biology studies have been used to complement these models to understand the mechanism of action of the anti-inflammatory effect.

9.3.1 *MORINGA OLEIFERA* LEAVES

One of the first *in vivo* studies on *Moringa* leaves was made by Venkataswera Rao et al. (1999). In this work, the aqueous and ethanolic extracts of *Moringa* leaves were tested in albino rats to probe their anti-inflammatory activities. The carrageenan-induced edema model was used to compare the effect of both extracts (at 200 mg/kg dose) and correlate the results with the ibuprofen activity (100 mg/kg). The extracts and drug were administered 30 min before the carrageenan injection to measure the percentage inhibition of edema at 1, 2, and 3 h. The administration of extracts showed

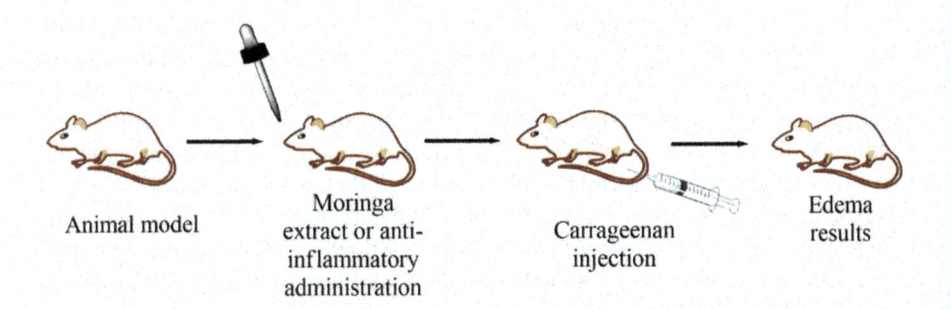

| Animal model | Moringa extract or anti-inflammatory administration | Carrageenan injection | Edema results |

FIGURE 9.3 The carrageenan-induced edema model.

the largest edema inhibition at 3 h of carrageenan injection and an anti-inflammatory effect like ibuprofen at 60 min of carrageenan injection, with an edema reduction of 48.2% (aqueous), 29.6% (ethanolic), and 53.7% (ibuprofen).

Anti-inflammatory activity in *Moringa* leaves has been studied using an *in vivo* model of acute inflammation by carrageenan-induced rat hind paw edema in Wistar albino rats (Sharma and Vaghela 2012). The edema was induced by injecting the sub-plantar region of the left hind paw with 0.1 ml of 1% w/v of carrageenan saline solution and following the progression of edema by increasing volume. The 50% ethanolic extract of *Moringa* leaves was administrated orally with 20, 200, and 500 mg/kg doses. The results of this study show a dose-dependent decrease of edema, having a maximum value at 3 h. In a similar study, the use of the ethanolic extract in an albino mice model showed that the anti-inflammatory activity was dose-dependent, decreasing the size of the induced edema in this animal model (Sharma and Vaghela 2012). Similar studies were made by Bhattacharya and colleagues, where the effect of ethanolic extract of leaves was tested in albino mice (Bhattacharya et al. 2014a) and albino rats (Bhattacharya et al. 2014b) animal models, showing a significant reduction of inflammation beside an improved analgesic activity.

In a rat model of pre-eclampsia, the ethanolic extract of *Moringa* leaves prevented the increase of IL-17 better than a low dose of aspirin (Batmomolin et al. 2020). In a different application of the anti-inflammatory properties of *Moringa* leaves, McKnight et al. (2014) proved that the ingestion of aqueous extracts of dried *Moringa* leaves in the form of brewed tea prevented acute lung inflammation. Acute lung inflammation was induced by inhalation of LPS (to cause a neutrophil-rich inflammatory response) or from dust collected from a swine confinement facility (DE – to cause cellular inflammation). Mice that consumed *Moringa* tea had significantly less protein and cellular influx into their lungs and presented more neutrophils with apoptotic morphology (McKnight et al. 2014).

In the study presented by Jurairat et al. (2012), *M. oleifera* leaf extract was evaluated to treat neuropathic pain. Male Wistar rats with diabetes mellitus were induced with neuropathic pain by damage in the sciatic nerve and administrated with the extract of *M. oleifera* leaves at doses of 100, 200, and 300 mg/kg daily for 21 days. The plant extract's analgesic effect was evaluated using Von Frey filament and hot plate tests every 3 days. The levels of oxidative damage markers, including malondialdehyde (MDA), superoxide dismutase (SOD), catalase (CAT), and glutathion peroxide (GSHPx) were evaluated to the injured sciatic nerve. The rats administrated with doses of 100 and 200 mg/kg of the extract reversed the decreased withdrawal threshold intensity and withdrawal latency in Von Frey filament and hot plate tests, respectively. The rats subjected to the 200 mg/kg dose of the extract also reversed the decreased activities of SOD and GSHPx and the elevation of MDA level in the injured nerve. This study suggests that an extract of *M. oleifera* leaves can attenuate neuropathic pain in diabetic conditions.

Besides flavonoids, leaves contain α-tocopherol, and when a *Moringa* leaf ethanolic extract was fractionated with different solvents, the dichloromethane fraction appeared to reduce the paw edema induced with carrageenan in mice more significantly than other fractions (Mahdi et al. 2017). A dichloromethane fraction had higher amounts of α-tocopherol, another important phytochemical with anti-inflammatory properties (Wallert et al. 2019).

9.3.2 *MORINGA OLEIFERA* BARK/WOOD/STEM

The effect of methanol stem extract on glycerol-induced acute kidney injury was studied by Adepapo et al. (2020). In this work, glycerol (50% v/v in sterile saline) was administrated intramuscularly in adult male Wistar rats to induce acute kidney injury. Using this model, the KIM-1 and nuclear factor kappa beta (NF-kB) membrane proteins signaling pathways can be monitored. In this study, five groups of animals were administrated as follows: group A (control) – distilled water; group B (toxicant) – glycerol alone on the 8th day; groups C and D were given with 50 mg/kg and 100 mg/kg of methanolic extract, respectively, for 7 days and glycerol on the 8th day; group E received 100 mg/kg of methanolic extract alone for 7 days and normal saline on the 8th day. Serum blood urea nitrogen (BUN), myeloperoxidase, creatinine, advanced oxidative products, malondialdehyde, reduced glutathione, superoxide dismutase, and protein carbonyl levels were measured to assess renal damage and possible ameliorative effects of the extract. Also, histopathological analysis of kidney tissues and immunohistochemical analysis of KIM-1 and NF-κB expression were carried out. The results showed decreased levels in renal enzymatic and non-enzymatic antioxidant defense system parameters as SOD, GPx (glutathione peroxide), GST, reduced glutathione, protein thiols, and non-protein thiol levels in the glycerol alone group compared with the other groups. Interestingly, the methanolic extract of *M. oleifera* showed higher inhibition of carrageenan-induced rat paw edema compared with the petroleum ether extract (Kumbhare and Sivakumar 2011).

9.3.3 *MORINGA OLEIFERA* SEEDS AND ROOTS

Hot water infusions of *M. oleifera* seeds and roots were tested by Cáceres et al. (1992) using carrageenan-induced edema in albino male rats. The administration of an aqueous extract of seeds showed greater activity than 50 mg/kg of phenylbutazone when tested at 1000 mg/kg dose. Among the first studies made with a methanolic extract of *Moringa* root was the work carried out by Ezeamuzie et al. (1996), where they proved that the anti-inflammatory effect shown by this extract was comparable with the effect of indomethacin at 5 mg/kg in the carrageenan-induced edema model using Wistar rats. The IC_{50} in swelling paw edema was 660 mg/kg. However, in the acute inflammation test, it was found that on day 6, the power of the extract was higher, with an IC_{50} of 302 mg/kg for the inhibition of cellular accumulation and of 315.5 mg/kg for fluid exudation. The maximum inhibition was obtained at a concentration of 600 mg/kg with 83.8% and 80.0%, respectively. In the chronic inflammation test, the extract remained effective but to a lesser degree than in acute inflammation.

On the other hand, Ndiaye et al. (2020) studied the anti-inflammatory effect of the aqueous extract of *Moringa* root and compared the effect of indomethacin administration (10 mg/kg) using the same carrageenan animal model. Here, they found that 750 mg/kg of aqueous extract administrated at 1, 3, and 5 h after carrageenan injection significantly inhibited the development of edema, presenting a reduction of 53.5%, 44.6%, and 51.1%, respectively.

9.3.4 MORINGA OLEIFERA FLOWERS

A hydroalcoholic extract of *M. oleifera* flowers has been studied to evaluate the anti-inflammatory effect and the arthritic index in Wistar rats using the Complete Freund's Adjuvant-induced arthritis in rats' model, molecular biology markers, and histopathological analysis (Mahajan et al. 2009). The effect of this extract was evaluated in an acute inflammatory process and arthritis disease as the chronic inflammation model obtained by *Mycobacterium tuberculosis* injection. The treatment with this extract in diseased rats showed that the paw edema and arthritic index were reduced; besides that, the rheumatoid factor (RF), and pro-inflammatory cytokines α-TNF and IL-1 also decreased at 200 mg/kg dose. The histopathological analysis showed less lymphocyte infiltration and less angiogenesis in animals treated with the extract compared with untreated ones.

9.4 CONCLUDING REMARKS

Although different *Moringa* parts have been consumed as food or used in traditional medicine, little is known about the phytochemicals responsible for anti-inflammatory effects. Leaves are rich sources of flavonoids such as quercetin and kaempferol, found as aglycones or glycosides. Even if these flavonoids have proven effects on the regulation of inflammation markers, other phytochemicals such as glucosinolates are highly relevant. Digestion is a crucial step that has only been included in certain experiments, showing that the resulting metabolites are more bioactive than those precursors found in the plant. Additionally, further research is needed to evaluate the effects of lipophilic compounds and the interactions involved in the anti-inflammatory effects observed *in vitro* and *in vivo* studies.

REFERENCES

Abd Rani, N.Z., Husain, K., and Kumolosasi, E. (2018). *Moringa* genus: a review of phytochemistry and pharmacology. *Frontiers in Pharmacology* 9: 108.

Adedapo, A.A., Etim, U., Falayi, O.O., Ogunpolu, B.S., Omobowale, T.O., Oyagbemi, A.A., et al. (2020). Methanol stem extract of Moringa oleifera mitigates glycerol-induced acute kidney damage in rats through modulation of KIM-1 and NF-kB signaling pathways. Scientific African 9:e00493.

Alhakmani, F., Kumar, S., and Khan, S.A. (2013). Estimation of total phenolic content, in vitro antioxidant and anti-inflammatory activity of flowers of *Moringa oleifera*. *Asian Pacific Journal of Tropical Biomedicine* 38), 623–627.

Batmomolin, A., Ahsan, A., Wiyasa, W.A. and Santoso, S. (2020). Ethanolic extract of Moringa oleifera leaves improve inflammation, angiogenesis, and blood pressure in rat model of preeclampsia. Journal of Applied Pharmaceutical Science 10(8): 52–57.

Beken, B., Serttas, R., Yazicioglu, M., Turkekul, K. and Erdogan, S. (2020). Quercetin improves inflammation, oxidative stress, and impaired wound healing in atopic dermatitis model of human keratinocytes. *Pediatric Allergy, Immunology, and Pulmonology* 33(2): 69–79.

Bhattacharya, A., Agrawal, D., Sahu, P.K., Kumar, S., Mishra, S.S., and Patnaik, S. (2014a). Analgesic effect of ethanolic leaf extract of *Moringa oleifera* on albino mice. *Indian Journal of Pain* 10: 89–94.

Bhattacharya, A., Agrawal, D., Sahu, P.K., Swain, T.R., Kumar, S., and Mishra, S.S. (2014b). Anti-inflammatory effect of ethanolic extract of *Moringa oleifera* leaves on albino rats. *Research Journal of Pharmaceutical, Biological and Chemical Sciences* 10: 540–544.

Bhattacharya, A., Tiwari, P., Sahu, P.K., and Kumar, S. (2018). A review of the phytochemical and pharmacological characteristics of *Moringa oleifera*. *Journal of Pharmacy and Bioallied Sciences* 10(4): 181–191.

Cáceres, A., Saravia, A., Rizzo, S., Zabala, L., De Leon, E., and Nave, F. (1992). Pharmacologic properties of *Moringa oleifera*. 2: Screening for antispasmodic, anti-inflammatory and diuretic activity. *Journal of Ethopharmacology* 36: 233–237.

Cheng, Y., Chen, Y., Li, J., Qu, H., Zhao, Y., Wen, C., et al. (2020). Dietary β-sitosterol regulates serum lipid level and improves immune function, antioxidant status, and intestinal morphology in broilers. *Poultry Science* 99(3): 1400–1408.

Cuellar-Núñez, M.L., Loarca-Pina, G., Berhow, M., and de Mejia, E.G. (2020). Glucosinolate-rich hydrolyzed extract from *Moringa oleifera* leaves decreased the production of TNF-α and IL-1β cytokines and induced ROS and apoptosis in human colon cancer cells. *Journal of Functional Foods* 75: 104270.

Cui, C., Chen, S., Wang, X., Yuan, G., Jiang, F., Chen, X., et al. (2019). Characterization of *Moringa oleifera* roots polysaccharide MRP-1 with anti-inflammatory effect. *International Journal of Biological Macromolecules* 132: 844–851.

Ding, K., Tan, Y.Y., Ding, Y., Fang, Y., Yang, X., Fang, J., et al. (2019). β-Sitosterol improves experimental colitis in mice with a target against pathogenic bacteria. *Journal of Cellular Biochemistry* 120(4): 5687–5694.

Ezeamuzie, I.C., Ambakederemo, A.W., Shode, F.O., and Ekwebelem, S.C. (1996). Anti-inflammatory effects of *Moringa oleifera* root extract. *International Journal of Pharmacognosy* 10: 207–212.

Fombang, E.N., Nobossé, P., Mbofung, C.M., and Singh, D. (2020). Optimising extraction of antioxidants from roasted *Moringa oleifera* Lam. leaves using response surface methodology. *Journal of Food Processing and Preservation* 44(6): e14482.

Fu, Z., Tang, S.F. and Hou, X. (2020). Probing the molecular toxic mechanism of di-(2-ethylhexyl) phthalate with glutathione transferase Phi8 from Arabidopsis thaliana. International Journal of Biological Macromolecules 145(15): 165–172.

Jurairat, K., Jintanaporn, W., Supaporn, M., Wipawee, T., Cholathip, T., Panakaporn, W., et al. (2012). *Moringa oleifera* leaves extract attenuates neuropathic pain induced by chronic constriction injury. *American Journal of Applied Sciences* 10: 1182–1187.

Koc, K., Geyikoglu, F., Cakmak, O., Koca, A., Kutlu, Z., Aysin, F., et al. (2021). The targets of β-sitosterol as a novel therapeutic against cardio-renal complications in acute renal ischemia/reperfusion damage. *Naunyn-Schmiedeberg's Archives of Pharmacology* 394: 469–479.

Kou, X., Li, B., Olayanju, J.B., Drake, J.M., and Chen, N. (2018). Nutraceutical or pharmacological potential of *Moringa oleifera* Lam. *Nutrients* 10(3): 343.

Kumar, S. (2017). Medicinal importance of *Moringa oleifera*: drumstick plant. *Indian Journal of Scientific Research* 10: 129–132.

Kumbhare, M. and Sivakumar, T. (2011). Anti-inflammatory and analgesic activity of stem bark of *Moringa oleifera*. *Pharmacology Online* 10: 641–650.

Kurano, M., Hasegawa, K., Kunimi, M., Hara, M., Yatomi, Y., Teramoto, T., et al. (2018.).Sitosterol prevents obesity-related chronic inflammation. *Biochimica et Biophysica Acta – Molecular and Cell Biology of Lipids* 1863(2): 191–198.

Leone, A., Spada, A., Battezzati, A., Schiraldi, A., Aristil, J., and Bertoli, S. (2015). Cultivation, genetic, ethnopharmacology, phytochemistry and pharmacology of *Moringa oleifera* leaves: an overview. *International Journal of Molecular Sciences* 16(6): 12791–12835.

Liao, P.-G., Lai, M.-H., Hsu, K.-P., Kuo, Y.-H., Chen, J., Tsai, M.-C., et al. (2018). Identification of β-sitosterol as *in vitro* anti-inflammatory constituent in *Moringa oleifera*. *Journal of Agricultural and Food Chemistry* 66(41): 10748–10759.

Liu, R., Hao, D., Xu, W., Li, J., Li, X., Shen, D., et al. (2019). β-Sitosterol modulates macrophage polarization and attenuates rheumatoid inflammation in mice. *Pharmaceutical Biology* 57(1): 161–168.

Mahajan, S.E. and Mehta, A.A. (2009). Anti-arthritic activity of hydroalcoholic extract of flowers of *Moringa oleifera* Lam. in Wistar rats. *Journal of Herbs, Spices and Medicinal Plants* 15: 149–163.

McKnight, M., Allen, J., Waterman, J.D., Hurley, S., Idassi, J., and Minor, R.C. (2014). Moringa tea blocks acute lung inflammation induced by swine confinement dust through a mechanism involving TNF-α expression, c-Jun N-terminal kinase activation and neutrophil regulation. *American Journal of Immunology* 10: 73–87.

Meireles, D., Gomes, J., Lopes, L., Hinzmann, M., and Machado, J. (2020). A review of properties, nutritional and pharmaceutical applications of *Moringa oleifera:* integrative approach on conventional and traditional Asian medicine. *Advances in Traditional Medicine* 20: 495–515.

Mishra, G., Singh, P., Verma, R., Kumar, S., Srivastav, S., Jha, K., et al. (2011). Traditional uses, phytochemistry and pharmacological properties of Moringa oleifera plant: An overview. Der Pharmacia Lettre 3: 141–164.

Molitorisova, M., Sutovska, M., Kazimierova, I., Barborikova, J., Joskova, M., Novakova, E., et al. (2021). The anti-asthmatic potential of flavonol kaempferol in an experimental model of allergic airway inflammation. *European Journal of Pharmacology* 891: 173698.

Ndiaye, M., Dieye, A.M., Mariko, F., Tall, A., Sall Diallo, A., and Faye, B. (2020). Contribution to the study of the anti-inflammatory activity of *Moringa oleifera* (Moringaceae). *Dakar Medical* 47: 210–212.

Sahakitpichan, P., Mahidol, C., Disadee, W., Ruchirawat, S. and Kanchanapoom, T. (2011). Unusual glycosides of pyrrole alkaloid and 4′-hydroxyphenylethanamide from leaves of Moringa oleifera. Phytochemistry 72(8): 791–795.

Saini, R.K., Sivanesan, I., and Keum, Y.-S. (2016). Phytochemicals of *Moringa oleifera*: a review of their nutritional, therapeutic and industrial significance. *3 Biotech* 6: 203.

Saleem, A., Saleem, M. and Akhtar, M.F. (2020). Antioxidant, anti-inflammatory and antiarthritic potential of *Moringa oleifera* Lam: an ethnomedicinal plant of Moringaceae family. *South African Journal of Botany* 128: 246–256.

Shallangwa, G.A., Dallatu, Y.A., Abechi, S.E., and Shuabu, H.U. (2016). *In vitro* anti-inflammatory evaluation of ethanol extracts of *Moringa oleifera, Thymus vulgaris* and their 1:1 extract blend on protein denaturation. *Trends in Science and Technology Journal* 1(2): 436–441.

Sharifudin, S.A., Fakurazi, S., Hidayat, M.T., Hairuszah, I., Aris Mohd Moklas, M., and Arulselvan, P. (2013). Therapeutic potential of *Moringa oleifera* extracts against acetaminophen-induced hepatotoxicity in rats. *Pharmaceutical Biology* 51(3): 279–288.

Sharma, R. and Vaghela, J.S. (2011). Anti-inflammatory activity of *Moringa oleifera* leaf and pod extracts against carrageenan induced paw edema in albino mice. *Pharmacology Online* 10: 140–144.

Srivastava, M., Dhakad, P.K., and Srivastava, B. (2020). A review on medicinal constituents and therapeutic potential of *Moringa oleifera*. *Universal Journal of Plant Science* 8(2): 22–33.

Venkataswera Rao, K.N., Gopalakrishnan, V., and Loganathan, V. (1999). Anti-inflammatory action of *Moringa oleifera* Lam. *Ancience Science of Life* 10: 195–198.

Wallert, M., Ziegler, M., Wang, X., Maluenda, A., Xu, X., Yap, M.L., et al. (2019). α-Tocopherol preserves cardiac function by reducing oxidative stress and inflammation in ischemia/reperfusion injury. *Redox Biology* 26: 101292.

Wang, J., Mao, J., Wang, R., Li, S., Wu, B. and Yuan, Y. (2020a). Kaempferol protects against cerebral ischemia reperfusion injury through intervening oxidative and inflammatory stress induced apoptosis. *Frontiers in Pharmacology* 11: 424.

Wang, Z., Sun, W., Sun, X., Wang, Y. and Zhou, M. (2020b). Kaempferol ameliorates cisplatin induced nephrotoxicity by modulating oxidative stress, inflammation and apoptosis via ERK and NF-κB pathways. *AMB Express* 10(1): 1–11.

Yan, G., Liping, S. and Yongliang, Z. (2020). UPLC-Q-Orbitrap-MS2 analysis of *Moringa oleifera* leaf extract and its antioxidant, antibacterial and anti-inflammatory activities. *Natural Product Research* 34(14): 2090–2094.

Yang, C., Yang, W., He, Z., He, H., Yang, X., Lu, Y., et al. (2020). Kaempferol improves lung ischemia–reperfusion injury via antiinflammation and antioxidative stress regulated by SIRT1/HMGB1/NF-κB Axis. *Frontiers in Pharmacology* 10: 1635.

Yao, H., Sun, J., Wei, J., Zhang, X., Chen, B. and Lin, Y. (2020). Kaempferol protects blood vessels from damage induced by oxidative stress and inflammation in association with the Nrf2/HO-1 signaling pathway. *Frontiers in Pharmacology* 11: 1118.

Yuan, K., Zhu, Q., Lu, Q., Jiang, H., Zhu, M., Li, X., et al. (2020). Quercetin alleviates rheumatoid arthritis by inhibiting neutrophil inflammatory activities. *The Journal of Nutritional Biochemistry* 84: 108454.

Zhang, Y., Peng, L., Li, W., Dai, T., Nie, L., Xie, J., et al. (2020). Polyphenol extract of *Moringa oleifera* leaves alleviates colonic inflammation in dextran sulfate sodium-treated mice. *Evidence-Based Complementary and Alternative Medicine* 2020: 6295402.

Zhou, B.X., Li, J., Liang, X.L., Pan, X.P., Hao, Y.B., Xie, P.F., et al. (2020). β-Sitosterol ameliorates influenza A virus-induced pro-inflammatory response and acute lung injury in mice by disrupting the cross-talk between RIG-I and IFN/STAT signaling. *Acta Pharmacologica Sinica* 41(9): 1178–1196.

10 Antidiabetic and Anti-obesity Properties of *Moringa* Species

Manuel A. Picos-Salas
Centro de Investigación en Alimentación y Desarrollo A.C., Carretera a Eldorado Km. 5.5, Col Campo El Diez, Culiacán, Sinaloa, 80110 México.

Luis A. Montoya-Inzunza
Centro de Investigación en Alimentación y Desarrollo A.C., Carretera a Eldorado Km. 5.5, Col Campo El Diez, Culiacán, Sinaloa, 80110 México.

Cristina Alicia Elizalde-Romero
Centro de Investigación en Alimentación y Desarrollo A.C., Carretera a Eldorado Km. 5.5, Col Campo El Diez, Culiacán, Sinaloa, 80110 México.

Erick P. Gutiérrez-Grijalva
Cátedras CONACYT-Centro de Investigación en Alimentación y Desarrollo A.C., Carretera a Eldorado Km. 5.5, Col Campo El Diez, Culiacán, Sinaloa, 80110 México.
Corresponding author: erick.gutierrez@ciad.mx

CONTENTS

DOI: 10.1201/9781003108863-10

10.1 DIABETES, OBESITY, AND THEIR IMPACT ON HEALTH

Diabetes mellitus, or simply diabetes, is a chronic metabolic disease related to high glucose levels in the blood, caused by null or low production of insulin in the pancreas or a deficiency in the use of the produced insulin (WHO 2020). This condition can be classified etiologically into type 1 and type 2, but other kinds are also present. Type 1 diabetes is caused by deficient insulin production generated by an autoimmune disease, leading to the destruction of pancreatic β-cells, but other forms of this type do not have a known etiology. In contrast, type 2 diabetes mellitus (T2DM), the most common form of diabetes, accounting for 90% of all cases, is produced by a disorder related to insulin resistance and insulin secretion, in which either can be predominant (Magliano et al. 2015; International Diabetes Federation 2020). It is estimated that around 463 million adults (aged 20–70 years) had diabetes in 2019, and the projections conclude there will be 578 million by 2030 and 700 million by 2045; with China, India, and the United States being the countries with the most cases of diabetes in adults (International Diabetes Federation 2020).

Among the most common symptoms of diabetes are excessive urination and thirst, blurry vision, constant hunger, weight loss without trying, fatigue, very dry skin, numb or tingling feet or hands, slowly healing sores, and more infections than usual. However, for T2DM, the symptoms can be mild or even asymptomatic (CDC 2019). Over time, diabetes can cause severe complications, including leg amputation, vision loss, kidney failure, and nerve damage; furthermore, adults have a higher risk of dying of heart attack and stroke (WHO 2020).

On the other hand, obesity, along with overweight, is defined by the WHO as "abnormal or excessive fat accumulation that presents a health risk," and a person with a body mass index (BMI) higher than 30 is considered obese; moreover, its prevalence in 2016 was around 650 million adults, 13% of the worldwide adult population. This condition is caused by excessive consumption of energy-dense foods rich in sugars and fat, along with physical inactivity (WHO 2020b); however, genetic factors are also involved (Hruby and Hu 2015).

Both conditions are closely linked, as approximately 90% of people with T2DM are obese (Sheehan and Ulchaker 2012). The pathophysiological link between diabetes and obesity is mostly attributed to insulin deficiency and resistance. An important factor in insulin resistance caused by obesity is the elevated and sustained elevation of free fatty acid (FFA) levels in plasma, which are released by central abdominal fat (Verma and Hussain 2016). Also, visceral adipocytes induce inflammation by secretion of cytokines, such as interleukin-6 and necrosis factor-α, which block insulin actions in various tissues and liver signaling and cause insulin resistance (Ota 2014). Moreover, a decrease in adiponectin, an anti-inflammatory and insulin sensitivity enhancer hormone, is also induced (Forny-Germano et al. 2019; Wondmkun 2020).

10.2 THE *MORINGA* GENUS

Moringa is the sole genus of the family Moringaceae and consists of 13 accepted species, namely *M. arborea, M. borziana, M. concanensis, M. drouhardii, M. hildebrandtii, M. longituba, M. oleifera, M. ovalifolia, M. peregrina, M. pygmaea, M. rivae, M. ruspoliana,* and *M. stenopetala*, which are native to southwest Asia, northeast Africa, southwest Africa, and Madagascar, but have been introduced to other tropical places such as Latin America, the Caribbean, Florida, and southeast Asia. It has been used in traditional medicine to heal skin wounds, diarrhea, sore throat, asthma, and fever, among others.

Of these species, *M. oleifera* has been the most studied; for instance, the leaves contain all the essential amino acids, along with a high amount of vitamins A and C, calcium, and potassium (Mahmood et al. 2010), and around 18.3%–28.7% of protein and 6.9%–10.7% of lipids, while the seeds have 38.5% and 42.2%, respectively (Al Juhaimi et al. 2017; Guzmán-Maldonado et al. 2020). The seed oil content of *M. peregrina* is 49.2%, while the leaves have around 25.9% of protein; also, both contain vitamins A and C, calcium, and potassium in enough quantities to supply the recommended dietary daily doses (Asghari et al. 2015).

Seeds of *M. concanensis* show 38.8% of oil and 30% of protein, while the leaves contain 27.2% of protein (Manzoor et al. 2007; Olson et al. 2016). Moreover, for the species *M. stenopetala*, the leaves have 29.7% protein, and the seeds contain 41.4% of oil and 42.6% of protein (Seifu 2012; Olson et al. 2016). This genus has also been studied for numerous potential health benefits, including antioxidant (Dehshahri et al. 2012), anticancer (Balamurugan et al. 2014; Mansour et al. 2019), anti-inflammatory (Tamrat et al. 2017; Albaayit et al. 2019), antidiabetic, and anti-obesity properties (Toma et al. 2014, 2015; Ullah et al. 2015; Hamed et al. 2020; Magaji et al. 2020), which are attributed to their bioactive phytochemical constituents.

10.3 ANTIDIABETIC PROPERTIES OF *MORINGA* SPECIES

10.3.1 IN VITRO ANTIDIABETIC STUDIES

10.3.1.1 α-Glucosidase Inhibition

Diabetes mellitus is mainly caused by a default in insulin activity and secretion, which results in a steady increase in serum glucose (hyperglycemia) and ends up causing various complications (Liu et al. 2013). A person's daily diet is provided by different macronutrients, such as carbohydrates, which end up being degraded to glucose, mostly by the enzymes α-glucosidase and α-amylase (Chen and Guo 2017). One way to control such glucose levels in this condition is through the controlled inhibition of these enzymes.

Some of the most commonly found phenolic compounds in plants are flavonoids, which have been shown to have great potential to inhibit various enzymes, including α-glucosidase. Some studies report inhibition of this enzyme by flavonoids. For instance, rutin, kaempferol acetyl glycoside, kaempferol 3-*O*-glucoside, quercetin-3-glucoside, and quercetin-3-acetyl-glucoside, found in *M. oleifera*, caused inhibition of 54.41% at a concentration of 100 µg/ml; also, an uncompetitive inhibition by the flavonoids present is suggested (Hamed et al. 2020). An IC_{50} of 123.34 µg/ml has also been reported in *M. oleifera* leaf ethanolic extract on this enzyme, attributed to its high total flavonoid content (Chen et al. 2020).

Studies using water as a solvent have achieved good results to inhibit this enzyme; an IC_{50} of 4.73 µg/ml was obtained using only distilled water for its leaf extract (Jimoh 2018); moreover, an EC_{50} of 40.7 µg/ml, 870.1 µg/ml, and 560.7 µg/ml have been reported for water extracts of leaves, main roots, and lateral roots, respectively; this activity is attributed to high antioxidant ability of the present compounds and the presence of numerous – OH groups that interact with the enzyme structure, altering it and preventing it from doing its physiological functions (Tshabalala et al. 2020); furthermore, leaf hydrophilic extracts where the most abundant compounds were phenolics, flavonoids, and condensed tannins presented an IC_{50} of 0.78 mg/ ml (Adisakwattana and Chanathong 2011). In the same way, the inhibitory effect of different types of extracts (water, methanol, hexane, and ethyl acetate) from different traditionally used parts of the plant (seed, leaves, and root) demonstrated that the root hexane extract had the best result when inhibiting this enzyme, with an IC_{50} of 0.382 mg/ml, thus demonstrating a better result than a commercial form of the treatment (acarbose; IC_{50} 0.884 mg/ml; Magaji et al. 2020). Furthermore, the inhibitory potential of *M. oleifera* zinc oxide nanoparticles on this enzyme has also been evaluated by obtaining an IC_{50} of 17.25 µg/ml (Rehana et al. 2017).

Different types of extracts affect the different parts of *Moringa*'s inhibitory capacity as do different types of sample processing, within which the drying method is important. Studies revealed that subjecting the sample to freeze-drying resulted in the best α-glucosidase inhibition (IC_{50} of 38.12 mg/ml) compared to other drying methods like sun, oven, and air drying; this is attributed to the concentration of phenolic compounds by the drying process, which can inhibit this enzyme in a dependent fashion (Ademiluyi et al. 2018).

10.3.1.2 α-Amylase Inhibition

The inhibitory properties of α-amylase by *Moringa* extracts have also been extensively evaluated. The hydroalcoholic leaf extract of *M. oleifera* showed an IC_{50} of 339.8 µg/ml, exhibiting high glycosides, oils, and polyphenolic compounds, where the latter show the greater inhibitory potential over the enzyme (Swamy and Meriga 2020). In other studies, *M. oleifera* leaf methanolic and hexane extracts have also shown great potential to inhibit this enzyme by obtaining an IC_{50} of 8.217 and 9.397 mg/ml, respectively (Magaji et al. 2020). The enzymatic inhibitory effect has also been reported testing other solvents such as water; by using an ultrasonic bath, the best phenolic compound extractions were obtained at temperatures of 60 and 80°C, having an IC_{50} of 0.540 g/ml in the extract of 80°C where large amounts of caffeoylquinic acid were detected (Hamed et al. 2020). Different processing methods such as sample drying affect this enzyme's inhibitory capacity; freeze-drying seems to have the best result in hydrophilic extracts of *M. oleifera* leaves compared to traditional methods, obtaining an IC_{50} of 64.29 mg/ml (Ademiluyi et al. 2018). The inhibitory activity presented by *M. oleifera* hydrophilic extracts is related to the presence of certain compounds that have already reported inhibitory properties in various enzymes, within which are the phenolic compounds: gallic acid, chlorogenic acid, ellagic acid, rutin, quercitrin, isoquercitin, quercetin, and kaempferol, with this extract having an IC_{50} of 6.49 µg/ml (Jimoh 2018).

Similarly, another study evaluated the inhibitory capacity of *M. oleifera* leaf hydrophilic extracts on the same enzyme. However, in this case, they measured the maximum inhibitory capacity of the extracts where inhibition of 80.5% was obtained at a concentration of 200 µg/ml and an IC_{50} of 52.5 µg/ml. This is related to the most-found flavonoid and phenolic compounds, thus attributing most of the hyperglycemic effect of such extracts to these types of compounds (Khan et al. 2017). Analyzing *M. peregrina*, the ability to inhibit the enzyme was similarly found with an IC_{50} of 1335.89 µg/ml (Ullah et al. 2015). Another way inhibition of this enzyme has been reported has been by zinc oxide nanoparticles of *M. oleifera* obtaining an IC_{50} of 35.72 µg/ml (Rehana et al. 2017). In this way, we can observe a great diversity of studies that support the various abilities of extracts of *Moringa* to inhibit such enzymes related to serum glucose levels.

10.3.1.3 Dipeptidyl Peptidase-4 Inhibition

Dipeptidyl peptidase-4 (DPP-4 or DPP-IV) is an important enzyme within T2DM; this is a serine protease, it works by degrading GLP-1 and GIP, which are metabolic hormones that promote insulin secretion. Therefore, one of the mechanisms for regulating the hyperglycemic state would be inhibiting the enzyme dipeptidyl peptidase, prolonging the hypoglycemic effects of these two hormones (Juillerat-Jeanneret 2014). Inhibitory capacity by *M. oleifera* hydrophilic extracts has been reported on this enzyme, obtaining an IC_{50} of 798 nM by a specific component, O-ethyl-4-[(α-L-rhamnosyloxy) benzyl] carbamate, reporting in various studies good pharmacokinetics and safety in various treatments (Yang et al. 2020). Despite this, the methanolic extracts from the entire *M. oleifera* plant have shown zero inhibition of this enzyme (Saidu et al. 2017). The discrepancy of this result with those previously reported shows the necessity of more studies of the behavior of *M. oleifera* extracts on this enzyme and the mechanism that these follow for inhibition. Although *M. oleifera* is the most studied species of this genus, the methanolic leaf extract of *M. peregrina* also exhibits the potential to inhibit this enzyme, with an IC_{50} of 1218.12 µg/ml; the same extract lengthened the activity of LPG-1 and GIP and its ability to regulate glucose in the body by stimulating insulin secretion (Ullah et al. 2015), in addition to also regulating the secretion of gastric acids to reduce the level of postpandrial glucose absorption (Drucker and Nauck 2006).

10.3.1.4 Other *In Vitro* Antidiabetic Studies

Another of the antidiabetic properties reported by ethanolic extracts of *M. stenopetala* leaves is its protective capacity from one of its major compounds, rutin, against oxidative stress exerted by H_2O_2, thus preventing the death of cell line β pancreatic 1,4E7 at a concentration of 200 µg/ml, demonstratingthat the *Moringa* plant has different properties that address the problem of diabetes (Habtemariam 2015). It has also been suggested that the antidiabetic properties of *Moringa* are related to inhibition of ATP-sensitive potassium channels in pancreatic cells, causing membrane depolarization affecting voltage-dependent calcium channels that end up opening up and cause an increase in intracellular calcium in beta cells and improving insulin release (Patel et al. 2017). As already mentioned, some of the most prevalent types of compounds in *Moringa* attributed to their antidiabetic ability are flavonoids, triterpenoids, sterols, phenols, and alkaloids, within which quercetin and kaempferol are the predominant flavonoids. At the same time, chlorogenic and quinic acids are the predominant phenolic acids (Mbikay 2012). The antidiabetic properties attributed to this type of compound report different mechanisms within which the protective ability over pancreatic cells induced with streptozotocin (STZ) stands out, as well as quercetin, preventing apoptosis (Panya et al. 2018). Glucose-6-phosphate translocase inhibition is also reported, and the way it reduces glycogenolysis and gluconeogenesis (Sneha and Kumar 2020).

10.4 *IN VIVO* ANTIDIABETIC STUDIES

As previously mentioned, *Moringa* plants are known for their anti-inflammatory, antioxidant, anticancer, and antidiabetic properties, and most of this research focuses on *M. oleifera* (Abd Rani et al. 2018). The main phytochemical groups commonly found in the *Moringa* genus are flavonoids, phenolic acids, and glucosinolates. Other natural compounds like terpenes, alkaloids, and sterols have also been reported. A summary of the recent *in vivo* reports of the antidiabetic and anti-obesity potential effects of *Moringa* plants is shown in Table 10.1.

The reported phytochemicals to which the antidiabetic and anti-obesity effect have been associated are 3-feruloylquinic acid, 4-*O*-caffeoylquinic acid, α-tocopherol acetate, hexadecenoic acid, kaempferol-3-glucoside, methyl heptadecanoate, niazirin, quercetin, phytol, quercetin-3-glucoside, and vitamin E (Figure 10.1). It is important to mention that for the mechanisms of action by which a phytochemical might exert their antidiabetic and anti-obesity activity they need to reach specific target places in tissues and organs. In this sense, the rule of 5 can be used to predict the drug-like passive cellular permeability and bioavailability of molecules (Lipinski et al. 1997; Q. Huang et al. 2020). Table 10.2 summarizes the predicted bioavailability of the most important identified molecules in *Moringa* species. Of the evaluated compounds, only 3-feruloylquinic acid, niazirin, and quercetin are bioavailable, so these molecules require further study to establish direct structure–activity relationships and isolation/synthesis of these molecules, aiming to develop new vehicles or administration strategies to exert a potential biopharmaceutical effect.

10.5 ANTI-OBESITY PROPERTIES OF *MORINGA* SPECIES

Obesity is a multifactorial health problem that often considers pharmacotherapy as an important element for its treatment. Several anti-obesity drugs have been withdrawn from the market due to their side effects, or continue to be available but lack treatment adherence from users. Regarding this issue, natural alternatives have been considered and evaluated in the search to reduce side effects. *Moringa* species have been evaluated to determine anti-obesity effects, with *in vitro* and *in vivo* evaluations (Nakayama et al. 2020).

10.5.1 *In Vitro* Anti-Obesity Studies

Anti-obesity effects can be evaluated *in vitro* using different assays such as pancreatic lipase inhibition, lipid peroxidation, anti-adipogenic effect, and inhibition of

TABLE 10.1
Antidiabetic properties of *Moringa* plants

Moringa species	Plant part	Treatment	Results	Compounds attributed to the antidiabetic effect	References
Moringa oleifera	Leaves	Ethanol extracts at a single daily dose of 200 mg/kg body weight were administered to adrenaline-induced hypertensive albino rats	*Moringa* extracts decreased blood glucose levels and serum triglyceride levels	Not reported	Ara et al. (2008)
M. oleifera	Leaves	Diabetic rats were treated with methanolic extracts of *M. oleifera* leaves at a dose of 300 mg/kg bodyweight for 60 days	The *Moringa* extracts treatment on diabetic rats increased plasma insulin and decreased serum glucose and glycated hemoglobin. On the other hand, *Moringa* increased the activities of antioxidant enzymes like superoxide dismutase, catalase, glutathione peroxidase, and glutathione-reductase	Hexadecenoic acid, phytol, α-tocopherol, heptadecanoic acid methyl ester, 11,14,17-eicosotrienoic acid methyl ester	Aju et al. (2019)
M. oleifera	Seeds	Niazirin, an *M. oleifera* seed compound, was administered daily to diabetic C57BL/6J mice at 5, 10 and 20 ml/kg body weight	Niazirin reduced insulin resistance, improved hyperglycemia, carbohydrate, and lipid metabolism	Niazirin	Bao et al. (2020)
M. oleifera	Leaves	The effect of *M. oleifera* extracts at 500 mg/kg body weight on diabetic rats was evaluated	*M. oleifera* showed no change in insulin secretion *in vivo*. However, it reduced serum glucose levels and retarded glucose absorption and inhibited α-amylase activity	Not reported	Bin Azad et al. (2017)

(continued)

TABLE 10.1 (Continued)
Antidiabetic properties of Moringa plants

Moringa species	Plant part	Treatment	Results	Compounds attributed to the antidiabetic effect	References
M. oleifera	Leaves	Ethanol extracts of *M. oleifera* leaves were administered to high-fat induced obesity rats at concentrations of 200 and 400 mg/kg body weight orally for 1 month	*Moringa* extracts at 400 mg/kg body weight reduced serum glucose and serum insulin nearly to basal levels. The extracts also reduced insulin resistance	Quercetin-*O*-rhamnosyl-hexosyl, kaempferol-*O*-xylosyl-apiosyl-acetyl, kaempferol-*O*-hexoside, feruloylquinic acid, quercetin, and kaempferide	Ezzat et al. (2020)
M. oleifera	Leaves	Ethanol extracts and butanol fractions were administered orally daily for 14 days to streptozotocin-induced diabetic rats at different concentrations (1000 and 500 mg/kg body weight)	The treatment did not increase blood insulin levels. However, ethanol extracts at 500 and 1000 mg/kg, and butanol fraction (500 mg/kg) returned to basal levels the concentration of the chemokine MCP-1	The effect was attributed partially to quercetin 3-β-D-glucoside, 4-*O*-caffeoylquinic acid, and kaempferol-3-*O*-glucoside present in *Moringa* extracts	Irfan et al. (2020a)
M. oleifera	Seeds	Evaluation of the effect of *M. oleifera* seed ethanol extract treatment on insulin resistance biomarkers	The *in-silico* analysis showed that moringa compounds can improve insulin resistance by acting on key targets related to inflammation and insulin	Glycosidic isothiocyanates and glycosidic benzylamines	Q. Huang et al. (2020)
M. oleifera	Leaves	Oral administration of *M. oleifera* extracts during 30 days at a concentration of 1000 mg/kg body weight	Ethanol and aqueous extracts decreased blood glucose levels to 4.33–4.43 mmol/l, plasma insulin levels and the HOMA-IR index	Not reported	Irfan et al. (2020b)

M. oleifera	Seed	*Moringa* seed extract was incorporated into obese C57B1/6J male mice for 12 weeks	The treatment reduced body weight, adiposity, improved glucose tolerance	The authors did not evaluate the phytochemical composition of the sample. However, they attributed the effect to quercetin, chlorogenic acid and mainly to glucosinolate-1 and isothiocyanate-1	Jaja-Chimedza et al. (2018)
M. oleifera	Seed	*Moringa* seed oil was tested on Sprgue–Dawely rats at a concentration of 800 mg/kg body weight for 8 weeks	*Moringa* treatment increased antioxidant enzyme activity, reduced lipid peroxidation, and reduced the expression of inflammatory cytokines like iNOS, which are associated with inflammation-related comorbidities of diabetes and obesity	Not reported	Kilany et al. (2020)
M. oleifera	Leaves	*Moringa* ethanolic extracts were orally administered to streptozotocin-induced diabetic rats at concentrations of 500 and 250 mg/kg body weight	The treatment with moringa extracts reduced water intake in diabetic rats, reduced fasting blood glucose levels, improved fasting plasma insulin concentration. The effective extract concentration was 500 mg/kg body weight	Acetic acid, octyl ester, 2-pyrrolidinone, furan-2-carboxylic acid, malic acid, glycine, benzeneacetic acid, 4-hydroxy-ethyl ester, D-galactopyranoside	Muzumbukilwa et al. (2019)
M. oleifera	Leaves	The effect of *Moringa* methanolic leaf extracts at 400 mg/kg body weight was evaluated on fructose-induced obesity hepatic lipid accumulation	*Moringa* treatment prevented the fructose-induced elevation in hepatic lipid stores, but did not prevent the fructose-induced hypertriglyceridemia. The extracts are proposed as potential agents against non-alcoholic fatty liver disease	Not reported	Muhammad et al. (2020)

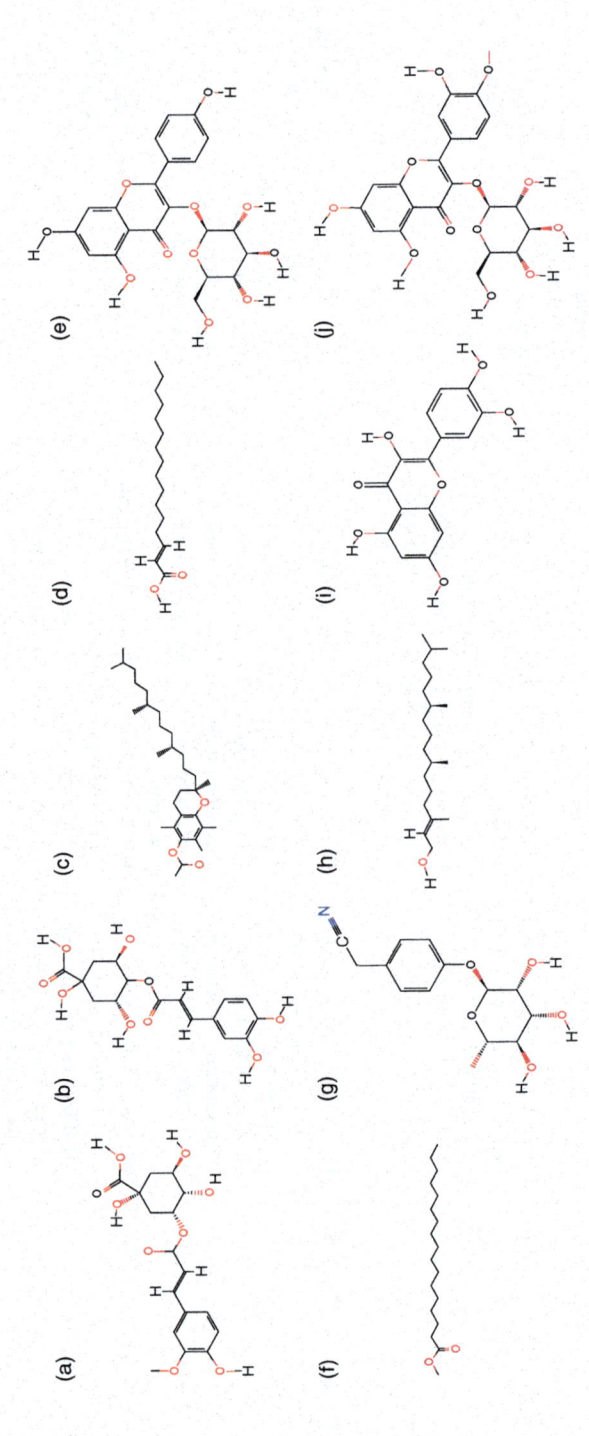

FIGURE 10.1 (a) 3-Feruoylquinic acid, (b) 4-*O*-caffeoylquinic acid, (c) α-tocopherol acetate, (d) hexadecenoic acid, (e) kaempferol-3-glucoside, (f) methyl heptadecanoate, (g) niazirin, (h) phytol, (i) quercetin.

TABLE 10.2
Prediction of the bioavailability of common phytochemicals in *Moringa* plants with antidiabetic and anti-obesity properties

Compound	Type of compound	Bioavailability Rule of 5
3-Feruloylquinic acid	Hydroxycinnamic acid	Yes
4-Caffeoylquinic acid	Hydroxycinnamic acid	No
α-Tocopherol acetate	Vitamin	No
Hexadecenoic acid	Monocarboxylic acid	No
Kaempferol-3-glucoside	Flavonol	No
Methyl heptadecanoate	Fatty acid methyl ester	No
Niazirin	Phenyl glycoside	Yes
Quercetin	Flavone (aglycone)	Yes
Phytol	Acyclic diterpenoid	No
Quercetin-3-glucoside	Flavone (glycosylated)	No
Vitamin E	Vitamin	No

lipogenesis, among others (Jakab et al. 2021). *Moringa oleifera* has been widely studied regarding this topic (Pagadala and Shankar 2020). Pancreatic lipase inhibition is a commonly used *in vitro* assay to evaluate potential anti-obesogenic materials. The assay may have slight modifications between one author and another. However, broadly, this evaluation uses a positive control (usually Orlistat), pancreatic lipase as the target enzyme, and a substrate to start the reaction, such as 4-nitrophenyl butyrate *p*-nitrophenyl palmitate (Lankatillake et al. 2021). Orlistat is a drug used to treat obesity due to its property of forming a covalent bond in the lipase active site (D.-S. Kim et al. 2020; S. Kim et al. 2020). Some secondary metabolites found in plants, such as caffeic acid, *p*-coumaric acid, and quercetin, have shown a lipase inhibition effect by binding in the active site through hydrogen bonds (D.-S. Kim et al. 2020).

Other anti-obesity effects can be determined by using the extracts or isolated compounds on 3T3-L1 preadipocytes. The cell differentiation, lipid accumulation, lipogenesis, viability, and other functional and structural aspects can be observed, measured, and compared to understand how some extracts and compounds have this potential as anti-obesogenic agents (Mopuri and Islam 2017).

Some researchers have evaluated antioxidant activity and anti-obesogenic and phenolic content of *M. oleifera* extracts by manipulating aspects of the extractions, such as solvent and temperature. It is interesting to observe and compare how the extracts' characteristics can change significantly with these manipulations. D.-S. Kim et al. (2020) evaluated aqueous and ethanolic extracts of *M. oleifera* from two different regions. When specific compounds were quantified using high-performance liquid chromatography (HPLC), it was observed that aqueous extracts were higher in gallic acid and caffeic acid, whereas rutin was higher in the ethanolic extracts. The region where they collected the samples was not as significant as the differences obtained while using different solvents. Antioxidant activity was also evaluated before the lipase inhibition assays; by ABTS radical scavenging and superoxide dismutase (SOD)-like activity. The antioxidant activity was higher when the extraction solvent was non-polar compared with a polar solvent. Results of lipase inhibition were also higher with a non-polar solvent (Table 10.3).

TABLE 10.3

In vitro **anti-obesogenic evaluation of *Moringa oleifera* extracts**

Material	Solvent	Concentration	Inhibition	Reference
Lipase inhibition				
Leaves (South Korea)	Ethanol	0.1 mg/ml	20.91%	D.-S. Kim et al. (2020)
Leaves (Cambodia)			32.58%	
Leaves (South Korea)	Water		10.63%	
Leaves (Cambodia)			18.95%	
Seeds (Nigeria)	Methanol	0.005 mg/ml	36.00%	Ajayi et al. (2020)
Young leaves (Kenia)	Ethanol	–	IC_{50} 0.18 mg/ml	G. L. Chen et al. (2020)
Leaves (India)	Hexane	–	IC_{50} 1.10 mg/ml	Swamy and Meriga (2020)
	Ethyl acetate		IC_{50} 0.60 mg/ml	
	Hydroalcoholic		IC_{50} 0.43 mg/ml	
	Aqueous		IC_{50} 0.82 mg/ml	
Leaves (China)	Water (80°C)	–	0.00017 mg/ml	Hamed et al. (2020)
Old and young leaves (India)	Sodium hydroxide	50 mg/ml	24.00%	Parwani (2016)
Effect in 3T3-L1 cells				
Leaves (Cambodia)	Water	0.025–0.10 mg/ml	Inhibition of adipogenesis and lipogenesis by inhibiting PPARγ,[1] FAS,[2] and ACC[3] markers	D.-S. Kim et al. (2020)
Leaves (China)	Petroleum ether	0.025–0.400 mg/ml	Inhibits lipogenesis by activating the AMPK[4] signaling pathway. Regulates the expression of adipogenesis-associated proteins and lipolysis-associated proteins	Xie et al. (2018)

Leaves	Ethanol	0.001 mg/ml	Adipogenesis reduction 60%	Nakayama et al. (2020)
Astragalin isolated from leaves	Hydroalcoholic	0.020 mg/ml	Reduces triglycerides content and lipid accumulation. Enhances glycerol release. Promotes lipolysis. Enhances the mRNA expression of adiponectin, and its secretion	Swamy et al. (2020)
Leaves (India)	Ethanol	IC_{50} 0.280 mg/ml	Suppresses adipogenesis markers such as PPARγ,[1] cEBPβ,[5] ADIPOR1,[6] and FABP4.[7] Decreases triglyceride content and promotes apoptosis of mature adipocytes	Balusamy et al. (2019)
Isolated benzyl isothiocyanate from peeled seeds (China)	Ethyl acetate and water (different fractions)	IC_{50} 0.0092 mg/ml	Inhibits lipid accumulation. Anti-adipogenic activity	L. H. Huang et al. (2020)
Leaves (Italy)	Ethanol	0.005–0.5 mg/ml	Induces thermogenic pathway. Reduces adipogenesis. Helps regulate adipokines secretion	Barbagallo et al. (2016)

[1] Peroxisome proliferator-activated receptor γ, [2]fatty acid synthesis, [3]acetyl-CoA carboxylase, [4]activated protein kinase, [5]enhancer-binding protein, [6]adiponectin receptor, [7]fatty-acid-binding proteins.

Hamed et al. (2020) published their results of lipase inhibition and antioxidant activity of aqueous extracts of *M. oleifera* manipulating water temperature during extraction; the extractions were made at 20, 40, 60, and 80°C. Total phenolic content and flavonoid content were superior in extractions performed at 60 and 80°C. This was also the case when evaluating antioxidant activity with ABST and DPPH radical scavenging, when lower IC_{50} was observed in both 60 and 80°C extracts, compared to the 20 and 40°C extracts. Besides, a similar behavior was observed when lipase inhibition potential was determined; the lower IC_{50}, and therefore the extract with better results, was the 80°C one (Table 10.3).

10.5.2 *In Vivo* Anti-Obesity Studies

Years of research on *M. oleifera* constituents, antioxidant activity, and potential health benefits have led to *in vivo* studies. Along with these studies, the reports of different communities using *Moringa* as part of their diet or as a supplement and natural remedy have been considered (Kou et al. 2018). The information available about this plant raises the possibility of studying its effect on live organisms. These studies allow researchers to evaluate effects, mechanisms of action, and possible applications, among other information about natural compounds and extracts (Table 10.4; Lakshmana Prabu et al. 2019).

Anti-obesogenic properties or effects are diverse, usually regarding weight loss, lipid accumulation, and lipid regulation. *Moringa oleifera* nutraceutical and pharmacological potential have been widely evaluated using *in vivo* models such as zebrafish, ducks, rats, humans, etc. (Mopuri and Islam 2017; Lakshmana Prabu et al. 2019).

Human studies suggest that *M. oleifera* supplementation could help obese patients regulate appetite and prevent common obesity-related diseases such as diabetes, cardiovascular diseases, and dyslipidemias. This gives *M. oleifera* components nutraceutical and pharmacological potential (Kou et al. 2018). The evidence shown for *in vitro* and *in vivo* evaluations of anti-obesogenic effects allows researchers to deepen the investigations in the *M. oleifera* mechanisms of action against obesity. In this regard, *M. oleifera* could be considered as an ingredient in nutraceutical products, a functional food, or if the active compounds are isolated, a pharmacologically active ingredient. Due to the numerous different effects that have been proved, there is a possibility to consider *M. oleifera* extracts or compounds as prevention or treatment, with patients presenting diagnoses of different severity and comorbidities caused by obesity.

10.6 CONCLUSIONS

In conclusion, extracts from the different *Moringa* species, mainly hydrophilic ones, where the predominant phytochemicals are flavonoids, triterpenoids, sterols, phenols, and alkaloids, possess promising antidiabetic properties such as inhibition of enzymes related to insulin secretion and carbohydrate digestion, as well as antioxidant capacity, protecting the body from one of the most common pathways of cell damage to the pancreas such as oxidative stress. The method of processing the sample is also important because subjecting it to different types of drying can increase or decrease the concentration of phytochemicals and consequently positively or negatively alter its antidiabetic properties.

TABLE 10.4
In vivo anti-obesogenic studies of *Moringa* extracts

Material	Model	Postulated function	Reference
Methanolic seed extract	Rats	Decrease in VLDLc[1] 100 mg/kg and 200 mg/kg Increase in HDLc[2] 200 mg/kg	Ajayi et al. (2020)
Fermented leaf dietary supplement	Ducks	Decrease lipid deposition in the liver and adipose tissues. Exerts an antilipogenic effect by downregulating the expression of hepatic FAS, ACC, and SREBP-1, along with the upregulation of lep-R	Zhang et al. (2020)
Isothiocyanates from aqueous leaf extract	Mice	Reduces in weight gain, hepatic adiposity, cholesterol levels, and inflammatory markers. Increases lipolysis	Waterman et al. (2015)
Leaves supplement	Guinea pigs	Suppresses hepatic lipid accumulation and inflammation	Almatrafi et al. (2017)
Ethanolic leaf extract	Rats	Downregulates mRNA expression of leptin and resistin. Upregulates adiponectin gene expression. Reduces body weight and improvement of the atherogenic and coronary index	Almatrafi et al. (2017)
Leaf alcohol extract	Rats	Reduces resistin levels and ameliorates dyslipidemia. Increases serum adiponectin levels	Ahmed et al. (2014)
Methanol leaf extracts	Rats	Decreases cholesterol levels in serum, kidney and liver	Bais et al. (2014)
Leaf powder supplement	Tilapia	Lowers lipid profile	El-Kassas et al. (2020)
Leaf powder supplement	Humans	Reduces total cholesterol, LDL, and TAG	Kumari (2010)
Cookies enriched with leaf powder	Humans	Reduces hunger ratings	Ahmad et al. (2018)
Leaf powder supplement	Zebrafish	Suppresses body weight increase	Nakayama et al. (2020)

[1] Very-low-density lipoprotein cholesterol, [2] high-density lipoprotein cholesterol.

REFERENCES

Abd Rani, N. Z., K. Husain, and E. Kumolosasi. (2018). *Moringa* genus: a review of phytochemistry and pharmacology. *Frontiers in Pharmacology* 9: 108. https://doi.org/10.3389/fphar.2018.00108.

Ademiluyi, A. O., O. H. Aladeselu, G. Oboh, and A. A. Boligon. (2018). Drying alters the phenolic constituents, antioxidant properties, alpha-amylase, and alpha-glucosidase inhibitory properties of moringa (*Moringa oleifera*) leaf. *Food Science & Nutrition* 6(8): 2123–2133. https://doi.org/10.1002/fsn3.770.

Adisakwattana, S., and B. Chanathong. (2011). Alpha-glucosidase inhibitory activity and lipid-lowering mechanisms of *Moringa oleifera* leaf extract. *European Review for Medical and Pharmacological Sciences* 15(7): 803–808.

Ahmad, J., I. Khan, S. K. Johnson, I. Alam, and Z. ud Din. (2018). Effect of incorporating *Stevia* and *Moringa* in cookies on postprandial glycemia, appetite, palatability, and gastrointestinal well-being. *Journal of the American College of Nutrition* 37(2): 133–139. https://doi.org/10.1080/07315724.2017.1372821.

Ahmed, H. H., F. M. Metwally, H. Rashad, A. M. Zaazaa, S. M. Ezzat, and M. M. Salama. (2014). *Moringa oleifera* offers a multi-mechanistic approach for management of obesity in rats. *International Journal of Pharmaceutical Sciences Review and Research* 29(2): 98–106.

Ajayi, T., M. Jones, O. J. Odumuwagun, and J. Olugbuyiro. (2020). Lipid altering potential of *Moringa oleifera* Lam seed extract and isolated constituents in Wistar rats. *African Journal Biomedical Research* 23: 77–85.

Aju, B. Y., R. Rajalakshmi, and S. Mini. (2019). Protective role of *Moringa oleifera* leaf extract on cardiac antioxidant status and lipid peroxidation in streptozotocin induced diabetic rats. *Heliyon* 5(12): 7. https://doi.org/10.1016/j.heliyon.2019.e02935.

Al Juhaimi, F., K. Ghafoor, E. E. Babiker, B. Matthäus, and M. M. Özcan. (2017). The biochemical composition of the leaves and seeds meals of *Moringa* species as non-conventional sources of nutrients. *Journal of Food Biochemistry* 41(1): e12322. https://doi.org/10.1111/jfbc.12322.

Albaayit, S. F. A., A. S. Kadhim Al-Khafaji, and H. S. Alnaimy. (2019). *In vitro* macrophage nitric oxide and interleukin-1 beta suppression by *Moringa peregrina* seed. *Turkish Journal of Pharmaceutical Sciences* 16(3): 362–365. https://doi.org/10.4274/tjps. galenos.2018.52244.

Almatrafi, M. M., M. Vergara-Jimenez, A. G. Murillo, G. H. Norris, C. N. Blesso, and M. L. Fernandez. (2017). *Moringa* leaves prevent hepatic lipid accumulation and inflammation in guinea pigs by reducing the expression of genes involved in lipid metabolism. *International Journal of Molecular Sciences* 18(7): 1330. https://doi.org/10.3390/ijms18071330. www.mdpi.com/1422-0067/18/7/1330.

Ara, N., M. Rashid, and M. S. Amran. (2008). Comparison of *Moringa oleifera* leaves extract with atenolol on serum triglyceride, serum cholesterol, blood glucose, heart weight, body weight in adrenaline induced rats. *Saudi Journal of Biological Sciences* 15(2): 253–258.

Asghari, G., A. Palizban, and B. Bakhshaei. (2015). Quantitative analysis of the nutritional components in leaves and seeds of the Persian *Moringa peregrina* (Forssk.) Fiori. *Pharmacognosy Research* 7(3): 242–248. https://doi.org/10.4103/0974-8490.157968.

Bais, S., G. S. Singh, and R. Sharma. (2014). Antiobesity and hypolipidemic activity of *Moringa oleifera* leaves against high fat diet-induced obesity in rats. *Advances in Biology* 2014: 162914. https://doi.org/10.1155/2014/162914.

Balamurugan, V., V. Balakrishnan, J. P. Robinson, and M. Ramakrishnan. (2014). Anticancer and apoptosis-inducing effects of *Moringa concanensis* using hepG2 cell lines. *Bangladesh Journal of Pharmacology* 9(4). https://dx.doi.org/10.3329/bjp.v9i4. 20481.

Balusamy, S. R., H. Perumalsamy, A. Ranjan, S. Park, and S. Ramani. (2019). A dietary vegetable, *Moringa oleifera* leaves (drumstick tree) induced fat cell apoptosis by inhibiting adipogenesis in 3T3-L1 adipocytes. *Journal of Functional Foods* 59: 251–260. https://doi.org/10.1016/j.jff.2019.05.029.

Bao, Y. F., J. B. Xiao, Z. B. Weng, X. Y. Lu, X. C. Shen, and F. Wang. (2020). A phenolic glycoside from *Moringa oleifera* Lam. improves the carbohydrate and lipid metabolisms through AMPK in db/db mice. *Food Chemistry* 311: 9. https://doi.org/10.1016/j.foodchem.2019.125948.

Barbagallo, I., L. Vanella, A. Distefano, D. Nicolosi, A. Maravigna, G. Lazzarino, et al. (2016). *Moringa oleifera* Lam. improves lipid metabolism during adipogenic differentiation of human stem cells. *European Review for Medical and Pharmacological Sciences* 20: 5223–5232.

Bin Azad, S., P. Ansari, S. Azam, S. M. Hossain, M. Ibtida-Bin Shahid, M. Hasan, et al. (2017). Anti-hyperglycaemic activity of *Moringa oleifera* is partly mediated by carbohydrase inhibition and glucose-fibre binding. *Bioscience Reports* 37: 11. https://doi.org/10.1042/bsr20170059.

CDC. (2019). Diabetes basics. Accessed August 1, 2021. www.cdc.gov/diabetes/basics/index.html.

Chen, G., and M. Guo. (2017). Rapid screening for α-glucosidase inhibitors from *Gymnema sylvestre* by affinity ultrafiltration–HPLC-MS. *Frontiers in Pharmacology* 8: 228. https://doi.org/10.3389/fphar.2017.00228.

Chen, G. L., Y. B. Xu, J. L. Wu, N. Li, and M. Q. Guo. (2020). Hypoglycemic and hypolipidemic effects of *Moringa oleifera* leaves and their functional chemical constituents. *Food Chemistry* 333: 127478. https://doi.org/10.1016/j.foodchem.2020.127478.

Dehshahri, S., M. Wink, S. Afsharypour, G. Asghari, and A. Mohaegheghzadeh. (2012). Antioxidant activity of methanolic leaf extract of *Moringa peregrina* (Forssk.) Fiori. *Research in Pharmaceutical Sciences* 7(2): 111–118.

Drucker, D. J., and M. A. Nauck. (2006). The incretin system: glucagon-like peptide-1 receptor agonists and dipeptidyl peptidase-4 inhibitors in type 2 diabetes. *Lancet* 368(9548): 1696–1705. https://doi.org/10.1016/s0140-6736(06)69705-5.

El-Kassas, S., S. E. Abdo, W. Abosheashaa, R. Mohamed, E. M. Moustafa, M. A. Helal, et al. (2020). Growth performance, serum lipid profile, intestinal morphometry, and growth and lipid indicator gene expression analysis of mono-sex Nile tilapia fed *Moringa oleifera* leaf powder. *Aquaculture Reports* 18: 100422. https://doi.org/10.1016/j.aqrep.2020.100422.

Ezzat, S. M., M. H. El Bishbishy, N. M. Aborehab, M. M. Salama, A. Hasheesh, A. A. Motaal, et al. (2020). Upregulation of MC4R and PPAR-alpha expression mediates the anti-obesity activity of *Moringa oleifera* Lam. in high-fat diet-induced obesity in rats. *Journal of Ethnopharmacology* 251: 11. https://doi.org/10.1016/j.jep.2020.112541.

Forny-Germano, L., F. G. De Felice, and M. Nunes Do Nascimento Vieira. (2019). The role of leptin and adiponectin in obesity-associated cognitive decline and Alzheimer's disease. *Frontiers in Neuroscience* 12. https://doi.org/10.3389/fnins.2018.01027.

Guzmán-Maldonado, S. H., M. J. López-Manzano, T. J. Madera-Santana, C. A. Núñez-Colín, C. P. Grijalva-Verdugo, A. G. Villa-Lerma, et al. (2020). Nutritional characterization of *Moringa olefiera* leaves, seeds, husks and flowers from two regions of Mexico. *Agronomia Colombiana* 38(2): 189–199.

Habtemariam, S. (2015). Investigation into the antioxidant and anti-diabetic potential of *Moringa stenopetala*: identification of the active principles. *Natural Product Communications* 10(3): 475–478.

Hamed, Y. S., M. Abdin, G. Chen, H. M. Saleem Akhtar, and X. Zeng. (2020). Effects of impregnate temperature on extraction of caffeoylquinic acid derivatives from *Moringa oleifera* leaves and evaluation of inhibitory activity on digestive enzyme, antioxidant, anti-proliferative and antibacterial activities of the extract. *International Journal of Food Science & Technology* 55(9): 3082–3090. https://doi.org/10.1111/ijfs.14572.

Hruby, A., and F. B. Hu. (2015). The epidemiology of obesity: a big picture. *PharmacoEconomics* 33(7): 673–689. https://doi.org/10.1007/s40273-014-0243-x.

Huang, L. H., C. M. Yuan, and Y. Wang. (2020). Bioactivity-guided identification of anti-adipogenic isothiocyanates in the moringa (*Moringa oleifera*) seed and investigation of the structure–activity relationship. *Molecules* 25(11): 9. https://doi.org/10.3390/molecules25112504.

Huang, Q., R. Liu, J. Liu, Q. Huang, S. Liu, and Y. P. Jiang. (2020). Integrated network pharmacology analysis and experimental validation to reveal the mechanism of anti-insulin resistance effects of *Moringa oleifera* seeds. *Drug Design Development and Therapy* 14: 4069–4084. https://doi.org/10.2147/dddt.S265198.

International Diabetes Federation. (2020). IDF diabetes atlas. Accessed August 1, 2021. www.diabetesatlas.org/.

Irfan, H. M., M. Z. Asmawi, N. A. K. Khan, S. B. Alamgeer, Z. Rasool, R. Shafiq ur, et al. (2020a). *In-vivo* and *in-vitro* studies to investigate the anti-diabetic mechanisms underlying *Moringa oleifera* leaf ethanol extracts. *Pakistan Journal of Pharmaceutical Sciences* 33(3): 1261–1270. https://doi.org/10.36721/pjps.2020.33.3.Sup.1261-1270.1.

Irfan, H. M., N. A. K. Khan, and M. Z. Asmawi. (2020b). *Moringa oleifera* Lam. leaf extracts reverse metabolic syndrome in Sprague Dawley rats fed high-fructose high fat diet for 60-days. *Archives of Physiology and Biochemistry*: 7. https://doi.org/10.1080/13813455.2020.1762661. <Go to ISI>://WOS:000534999500001.

Jaja-Chimedza, A., L. Zhang, K. Wolff, B. L. Graf, P. Kuhn, K. Moskal, et al. (2018). A dietary isothiocyanate-enriched moringa (*Moringa oleifera*) seed extract improves glucose tolerance in a high-fat-diet mouse model and modulates the gut microbiome. *Journal of Functional Foods* 47: 376–385. https://doi.org/10.1016/j.jff.2018.05.056.

Jakab, J., B. Miskic, S. Miksic, B. Juranic, V. Cosic, D. Schwarz, et al. (2021). Adipogenesis as a potential anti-obesity target: a review of pharmacological treatment and natural products. *Diabetes Metabolic Syndrome and Obesity-Targets and Therapy* 14: 67–83. https://doi.org/10.2147/dmso.s281186. <Go to ISI>://WOS:000607141800001.

Jimoh, T. O. (2018). Enzymes inhibitory and radical scavenging potentials of two selected tropical vegetable (*Moringa oleifera* and *Telfairia occidentalis*) leaves relevant to type 2 diabetes mellitus. *Revista Brasileira De Farmacognosia – Brazilian Journal of Pharmacognosy* 28(1): 73–79. https://doi.org/10.1016/j.bjp.2017.04.003.

Juillerat-Jeanneret, L. (2014). Dipeptidyl peptidase IV and its inhibitors: therapeutics for type 2 diabetes and what else? *Journal of Medicinal Chemistry* 57(6): 2197–2212. https://doi.org/10.1021/jm400658e.

Khan, W., R. Parveen, K. Chester, S. Parveen, and S. Ahmad. (2017). Hypoglycemic potential of aqueous extract of *Moringa oleifera* leaf and *in vivo* GC-MS metabolomics. *Frontiers in Pharmacology* 8: 577. https://doi.org/10.3389/fphar.2017.00577.

Kilany, O. E., H. M. A. Abdelrazek, T. S. Aldayel, S. Abdo, and M. M. A. Mahmoud. (2020). Anti-obesity potential of *Moringa olifera* seed extract and lycopene on high fat diet induced obesity in male Sprauge Dawely rats. *Saudi Journal of Biological Sciences* 27(10): 2733–2746. https://doi.org/10.1016/j.sjbs.2020.06.026.

Kim, D.-S., M.-H. Choi, and H.-J. Shin. (2020). Extracts of *Moringa oleifera* leaves from different cultivation regions show both antioxidant and anti-obesity activities. *Journal of Food Biochemistry* 44(7): e13282. https://doi.org/10.1111/jfbc.13282.

Kim, S., J. H. Kim, S. H. Seok, and E. S. Park. (2020). Anti-obesity effect with reduced adverse effect of the co-administration of mini-tablets containing orlistat and mini-tablets containing xanthan gum: *in vitro* and *in vivo* evaluation. *International Journal of Pharmaceutics* 591: 11. https://doi.org/10.1016/j.ijpharm.2020.119998.

Kou, X., B. Li, J. B. Olayanju, J. M. Drake, and N. Chen. (2018). Nutraceutical or pharmacological potential of *Moringa oleifera* Lam. *Nutrients* 10(3): 343. https://doi.org/10.3390/nu10030343. https://pubmed.ncbi.nlm.nih.gov/29534518.

Kumari, D. J. (2010). Hypoglycaemic effect of *Moringa oleifera* and *Azadirachta indica* in type 2 diabetes mellitus. *Bioscan* 5(20): 211–214.

Lakshmana Prabu, S., A. Umamaheswari, and A. Puratchikody. (2019). Phytopharmacological potential of the natural gift *Moringa oleifera* Lam and its therapeutic application: an overview. *Asian Pacific Journal of Tropical Medicine* 12(11): 485–498. https://doi.org/10.4103/1995-7645.271288.

Lankatillake, C., S. Q. Luo, M. Flavel, G. B. Lenon, H. Gill, T. Huynh, et al. (2021). Screening natural product extracts for potential enzyme inhibitors: protocols, and the standardisation of the usage of blanks in α-amylase, α-glucosidase and lipase assays. *Plant Methods* 17(1): 19. https://doi.org/10.1186/s13007-020-00702-5.

Lipinski, C. A., F. Lombardo, B. W. Dominy, and P. J. Feeney. (1997). Experimental and computational approaches to estimate solubility and permeability in drug discovery and development settings. *Advanced Drug Delivery Reviews* 23(1): 3–25. https://doi.org/10.1016/S0169-409X(96)00423-1.

Liu, S., D. Li, B. Huang, Y. Chen, X. Lu, and Y. Wang. (2013). Inhibition of pancreatic lipase, α-glucosidase, α-amylase, and hypolipidemic effects of the total flavonoids from *Nelumbo nucifera* leaves. *Journal of Ethnopharmacology* 149(1): 263–269. https://doi.org/10.1016/j.jep.2013.06.034.

Magaji, U. F., O. Sacan, and R. Yanardag. (2020). Alpha amylase, alpha glucosidase and glycation inhibitory activity of *Moringa oleifera* extracts. *South African Journal of Botany* 128: 225–230. https://doi.org/10.1016/j.sajb.2019.11.024.

Magliano, D. J., P. Zimmet, and J. E. Shaw. (2015). Classification of diabetes mellitus and other categories of glucose intolerance. In *International Textbook of Diabetes Mellitus*, edited by Ralph A. DeFronzo, Ele Ferrannini, Paul Zimmet and K. Geroge M. M. Alberti (pp. 3–16). Oxford: Wiley Blackwell.

Mahmood, K. T., T. Mugal, and I. Ul Haq. (2010). *Moringa oleifera*: a natural gift – a review. *Journal of Pharmaceutical Sciences and Research* 2(11): 775–781.

Mansour, M., M. F. Mohamed, A. Elhalwagi, H. A. El-Itriby, H. H. Shawki, and I. A. Abdelhamid. (2019). *Moringa peregrina* leaves extracts induce apoptosis and cell cycle arrest of hepatocellular carcinoma. *BioMed Research International* 2019: 2698570. https://doi.org/10.1155/2019/2698570.

Manzoor, M., F. Anwar, T. Iqbal, and M. I. Bhanger. (2007). Physico-chemical characterization of *Moringa concanensis* seeds and seed oil. *Journal of the American Oil Chemists' Society* 84(5): 413–419. https://doi.org/10.1007/s11746-007-1055-3.

Mbikay, M. (2012). Therapeutic potential of *Moringa oleifera* leaves in chronic hyperglycemia and dyslipidemia: a review. *Frontiers in Pharmacology* 3: 24. https://doi.org/10.1016/j.sajb.2019.11.024.

Mopuri, R. G., and M. Islam. (2017). Medicinal plants and phytochemicals with anti-obesogenic potentials: a review. *Biomedicine & Pharmacotherapy* 89: 1442–1452. https://doi.org/10.1016/j.biopha.2017.02.108.

Muhammad, N., K. G. Ibrahim, A. R. Ndhlala, and K. H. Erlwanger. (2020). *Moringa oleifera* Lam. prevents the development of high fructose diet-induced fatty liver. *South African Journal of Botany* 129: 32–39. https://doi.org/10.1016/j.sajb.2018.12.003.

Muzumbukilwa, W. T., M. Nlooto, and P. M. Oroma Owira. (2019). Hepatoprotective effects of *Moringa oleifera* Lam (Moringaceae) leaf extracts in streptozotocin-induced diabetes in rats. *Journal of Functional Foods* 57: 75–82. https://doi.org/10.1016/j.jff.2019.03.050.

Nakayama, H., K. Hata, I. Matsuoka, L. Zang, Y. Kim, D. Chu, et al. (2020). Anti-obesity natural products tested in juvenile zebrafish obesogenic tests and mouse 3T3-L1 adipogenesis assays. *Molecules* 25(24). https://doi.org/10.3390/molecules25245840.

Olson, M. E., R. P. Sankaran, J. W. Fahey, M. A. Grusak, D. Odee, and W. Nouman. (2016). Leaf protein and mineral concentrations across the "miracle tree" genus *Moringa*. *PLoS ONE* 11(7): e0159782. https://doi.org/10.1371/journal.pone.0159782.

Ota, T. (2014). Obesity-induced inflammation and insulin resistance. *Frontiers in Endocrinology* 5. https://doi.org/10.3389/fendo.2014.00204.

Pagadala, P., and V. Shankar. (2020). *Moringa olifera*: constituents and protective effects on organ systems. *Physiology and Pharmacology (Iran)* 24(2): 82–88. https://doi.org/10.32598/ppj.24.2.40.

Panya, T., N. Chansri, B. Sripanidkulchai, and S. Daodee. (2018). Additional antioxidants on the determination of quercetin from *Moringa oleifera* leaves and variation content from different sources. *International Food Research Journal* 25(1): 51–55.

Parwani, L. (2016). Effect of temperature on α-glucosidase, lipase inhibition activity and other nutritional properties of *Moringa oleifera* leaves: Intended to be used as daily anti-diabetic therapeutic food. *Journal of Food and Nutrition Research* 55.

Patel, A. B., D. D. Prajapati, and Y. Patel. (2017). Anti-diabetic activity of *Moringa oleifera* Lam. *Algerian Journal of Natural Products* 5(2): 446–453. https://doi.org/10.5281/zenodo.865484.

Rehana, D., D. Mahendiran, R. S. Kumar, and A. K. Rahiman. (2017). *In vitro* antioxidant and anti-diabetic activities of zinc oxide nanoparticles synthesized using different plant extracts. *Bioprocess and Biosystems Engineering* 40(6): 943–957. https://doi.org/10.1007/s00449-017-1758-2.

Saidu, Y., S. A. Muhammad, A. Y. Abbas, A. Onu, I. M. Tsado, and L. Muhammad. (2017). *In vitro* screening for protein tyrosine phosphatase 1B and dipeptidyl peptidase IV inhibitors from selected Nigerian medicinal plants. *Journal of Intercultural Ethnopharmacology* 6(2): 154–157. https://doi.org/10.5455/jice.20161219011346.

Seifu, E. (2012). Physicochemical properties of *Moringa stenopetala* (Haleko) seeds. *Journal of Biological Sciences* 12(3): 197–201. https://doi.org/10.3923/jbs.2012.197.201.

Sheehan, J. P., and M. M. Ulchaker. (2012). *Obesity and Type 2 Diabetes Mellitus.* New York: Oxford University Press.

Sneha, M. K., and S. Kumar. (2020). Efficacy and mechanism of action of *Moringa oleifera* in diabetes. *International Journal of Pharmaceutical Sciences and Research* 11(9): 4201–4213. https://doi.org/10.13040/ijpsr.0975-8232.11(9).4201-13.

Swamy, G. M., and B. Meriga. (2020). Therapeutic effect of *Moringa oleifera* leaf extracts on oxidative stress and key metabolic enzymes related to obesity. *International Journal of Pharmaceutical Sciences and Research* 11(4): 1949–1957. https://doi.org/10.13040/ijpsr.0975-8232.11(4).1949-57.

Swamy, G. M., G. Ramesh, R. D. Prasad, and B. Meriga. (2020). Astragalin, (3-*O*-glucoside of kaempferol), isolated from *Moringa oleifera* leaves modulates leptin, adiponectin secretion and inhibits adipogenesis in 3T3-L1 adipocytes. *Archives of Physiology and Biochemistry*: 7. https://doi.org/10.1080/13813455.2020.1740742.

Tamrat, Y., T. Nedi, S. Assefa, T. Teklehaymanot, and W. Shibeshi. (2017). Anti-inflammatory and analgesic activities of solvent fractions of the leaves of *Moringa stenopetala* Bak. (Moringaceae) in mice models. *BMC Complementary and Alternative Medicine* 17(1):473. https://doi.org/10.1186/s12906-017-1982-y.

Toma, A., E. Makonnen, Y. Mekonnen, A. Debella, and S. Addisakwattana. (2014). Intestinal α-glucosidase and some pancreatic enzymes inhibitory effect of hydroalcholic extract of *Moringa stenopetala* leaves. *BMC Complementary and Alternative Medicine* 14(1): 180. https://doi.org/10.1186/1472-6882-14-180.

Toma, A., E. Makonnen, Y. Mekonnen, A. Debella, and S. Adisakwattana. (2015). Anti-diabetic activities of aqueous ethanol and *n*-butanol fraction of *Moringa stenopetala* leaves in streptozotocin-induced diabetic rats. *BMC Complementary and Alternative Medicine* 15(1): 242. https://doi.org/10.1186/s12906-015-0779-0.

Tshabalala, T., A. R. Ndhlala, B. Ncube, H. A. Abdelgadir, and J. Van Staden. (2020). Potential substitution of the root with the leaf in the use of *Moringa oleifera* for antimicrobial, anti-diabetic and antioxidant properties. *South African Journal of Botany* 129: 106–112. https://doi.org/10.1016/j.sajb.2019.01.029.

Ullah, M. F., S. H. Bhat, and F. M. Abuduhier. (2015). Anti-diabetic potential of hydro-alcoholic extract of *Moringa peregrina* leaves: implication as functional food for prophylactic intervention in prediabetic stage. *Journal of Food Biochemistry* 39(4): 360–367.

Verma, S., and M. E. Hussain. (2016). Obesity and diabetes: an update. *Diabetes & Metabolic Syndrome: Clinical Research & Reviews* 11(1): 73–79. https://doi.org/10.1016/j.dsx.2016.06.017.

Waterman, C., P. Rojas-Silva, T. Tumer, P. Kuhn, A. Richard, S. Wicks, et al. (2015). Isothiocyanate-rich *Moringa oleifera* extract reduces weight gain, insulin resistance and hepatic gluconeogenesis in mice. *Molecular Nutrition & Food Research* 59: 1013–1024. https://doi.org/10.1002/mnfr.201400679.

WHO. (2020). Diabetes. Accessed August 1, 2021. www.who.int/news-room/fact-sheets/detail/diabetes.

Wondmkun, Y. T. (2020). Obesity, insulin resistance, and type 2 diabetes: associations and therapeutic implications. *Diabetes, Metabolic Syndrome and Obesity: Targets and Therapy* 13: 3611–3616. https://doi.org/10.2147/dmso.s275898.

Xie, J., Y. Wang, W.-W. Jiang, X.-F. Luo, T.-Y. Dai, L. Peng, et al. (2018). *Moringa oleifera* leaf petroleum ether extract inhibits lipogenesis by activating the AMPK signaling pathway. *Frontiers in Pharmacology* 9: 1447. https://doi.org/10.3389/fphar.2018.01447.

Yang, Y., C. Y. Shi, J. Xie, J. H. Dai, S. L. He, and Y. Tian. (2020). Identification of potential dipeptidyl peptidase (DPP)-IV Inhibitors among *Moringa oleifera* phytochemicals by virtual screening, molecular docking analysis, ADME/T-based prediction, and *in vitro* analyses. *Molecules* 25(1): 189. https://doi.org/10.3390/molecules25010189.

Zhang, X., Z. Sun, J. Cai, G. Wang, J. Wang, Z. Zhu, et al. (2020). Dietary supplementation with fermented *Moringa oleifera* leaves inhibits the lipogenesis in the liver of meat ducks. *Animal Feed Science and Technology* 260: 114336. https://doi.org/10.1016/j.anifeedsci.2019.114336.

11 Perspectives on the Study of *Moringa*

Erick P. Gutiérrez-Grijalva

Cátedras CONACYT-Centro de Investigación en Alimentación y Desarrollo A.C., Carretera a Eldorado Km. 5.5, Col Campo El Diez, Culiacán, Sinaloa, 80110 México.

J. Basilio Heredia

Centro de Investigación en Alimentación y Desarrollo A.C., Carretera a Eldorado Km. 5.5, Col Campo ElDiez, Culiacán, Sinaloa, 80110 México.
Corresponding author: jbheredia@ciad.mx

CONTENTS

11.1 INTRODUCTION

As stated in the chapters of this work, *Moringa* is a rich source of phytochemicals and natural products with potential biopharmaceutical applications. Alkaloids, phenolic compounds, flavonoids, some glycosides, among other compounds, have shown anti-inflammatory, antioxidant, anticancer, antidiabetic, and anti-obesity properties. Moreover, the most widely studied *Moringa* species, without a doubt, is *M. oleifera*, with 1810 and 2510 publications in Web of Science and Scopus databases, respectively (Figure 11.1). Interestingly, *Moringa* species originate in Africa and Asia; however, these plants have been distributed in many tropical countries. Moreover, its use has been popularized in many more countries.

The main research areas of interest in *Moringa* studies are agriculture, chemistry, food science and technology, pharmacology, medicine, engineering, and plant sciences (Figure 11.2). With regard to botanical and agronomical studies, Chapters 1 and 3 covered all related information. Many efforts have also been trying to implement plant breeding improvements, but limited to the *M. oleifera*. In Chapter 2, the authors discussed the genetic diversity and molecular markers commonly used for this purpose. Further studies are needed regarding the other *Moringa* species to improve plant breeding programs.

However, the most common area of interest of the *Moringa* uses and applications is its biological effects on human health. The most common causes of death worldwide

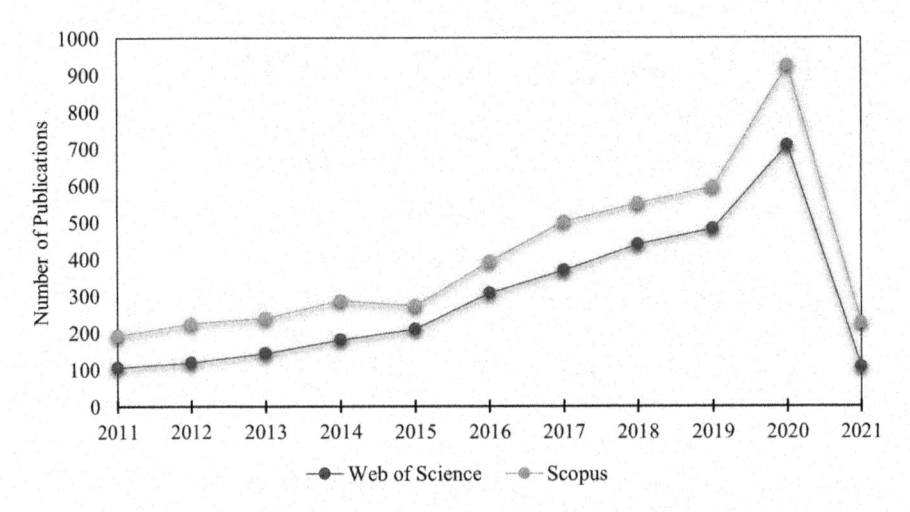

FIGURE 11.1 Recent publications with the subject "Moringa" in the last 10 years from the Scopus and Web of Science databases. There is an increasing trend to study *Moringa* for its many properties.

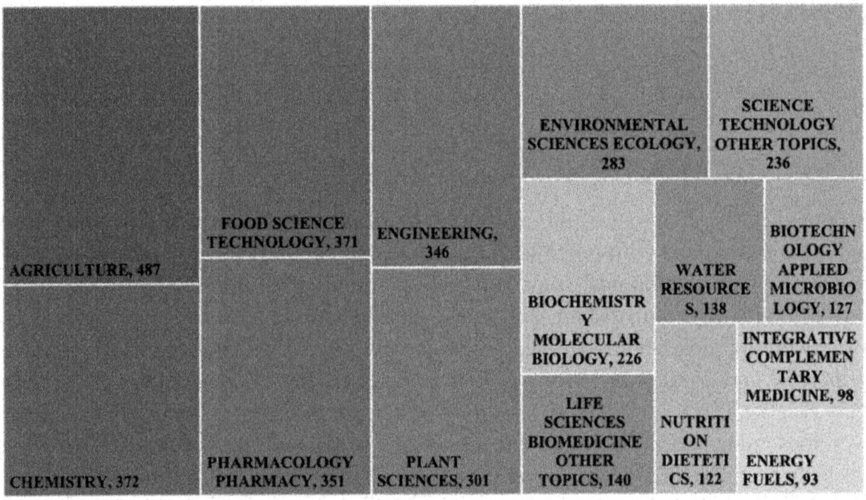

FIGURE 11.2 Publications regarding "Moringa" by subject area from 2011 to 2021 using the Web of Science and Scopus metrics.

are attributed to non-communicable diseases. On this subject, the World Health Organization has stated that non-communicable diseases kill nearly 41 million people per year, around the 70% of deaths altogether (World Health Organization 2018). The leading non-communicable diseases in terms of mortality are cardiovascular diseases, different types of cancer, and diabetes. They cause a direct impact on the global health and economy, which are related to premature deaths caused by these diseases.

Non-communicable diseases have been related to metabolic risk factors like raised blood pressure, overweight and obesity, hyperglycemia, hyperlipidemia, and lifestyle factors like tobacco use, alcohol abuse, unhealthy diet, and physical inactivity. So far, *Moringa* has been a promising alternative to conventional pharmacological drugs to treat these ailments.

In this work, we have assessed the main areas of interest in *Moringa* research; Chapters 6 to 10 focused on the phytochemicals found in *Moringa* species, their high protein content, the feasibility of using them as bioactive peptides, the antioxidant capacity of moringa compounds to act against oxidative stress, and the mechanisms by which they can delay the onset or the comorbidities of numerous non-communicable diseases (Halliwell and Gutteridge 2015).

Furthermore, some aspects of the ethnopharmacological studies have been pointed out by Gertsch (2009), Cos et al. (2006), Atanasov et al. (2021), and others to consider these studies of the scientific impact and ethnopharmacological significance. One of the most important factors strongly suggested to be considered in new studies is the relationship between the concentrations employed in different biochemical and pharmacological analyses. The authors state the "overinterpretation of *in vitro* data" (Gertsch 2009). Furthermore, it is stated that the lack of consideration of the physiological concentrations during the design of cellular-based *in vitro* assays can result in a "forced" positive result without scientific meaningfulness. In this work, the reviewed papers on the chapters of phytochemistry and bioactive effects of *Moringa* do not consider this aspect in their experiments. Moreover, as Gertsch (2009) mentioned, high concentrations of natural products can target multiple proteins and overestimate the *in vitro* bioactivity. Another aspect that is rarely (or never) considered is the possible synergistic or antagonistic effect of the compounds in an extract. For instance, few publications evaluate individually the standards of the main constituents hypothesized for an *in vitro* bioactive effect. New strategies need to be set following a meticulously designed ethnopharmacological evaluation (Figure 11.3).

Moreover, the bioavailability of the bioactive compounds distributed in the *Moringa* species is rarely given. Thus, the potential biopharmacological effect can be misleading towards non-bioavailable compounds. Furthermore, *in vitro* and *in vivo* assays are needed to elucidate those molecules that will be distributed in the plasma and target tissues and organs.

11.2 CONCLUSIONS

Further studies regarding *Moringa*, its phytochemicals and natural products, and their relationship with their potential anti-obesity, anti-inflammatory, antioxidant, antiproliferative, and other properties needs to be assessed following the suggested recommendations. The use of physiologically significant concentrations of extracts and molecules in *in vitro* studies, as well as the establishment of pharmacokinetic parameters, are also needed during preclinical and clinical studies. Moreover, the possible interactions between moringa compounds and target signaling proteins involved in their bioactive potential can also be evaluated incorporating *in silico* studies.

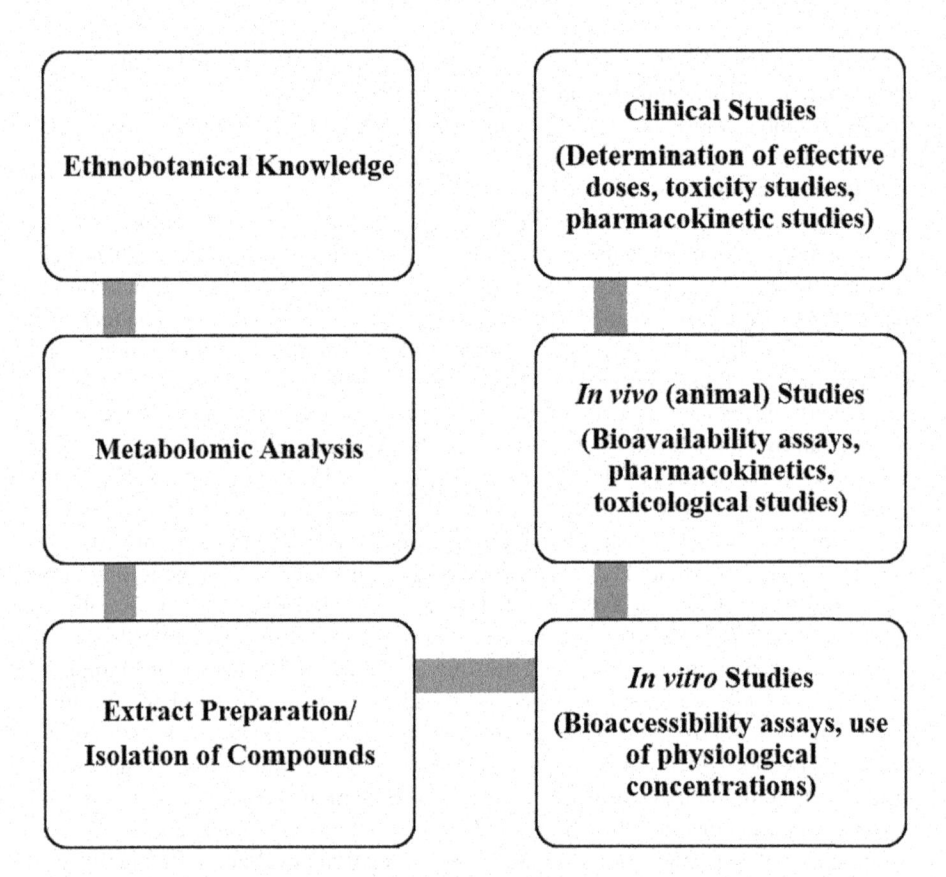

FIGURE 11.3 Proposed strategy during the design and development of ethnopharmacological studies.

REFERENCES

Atanasov, A. G., S. B. Zotchev, V. M. Dirsch, I. E. Orhan, M. Banach, J. M. Rollinger, et al. (2021). Natural products in drug discovery: advances and opportunities. *Nature Reviews Drug Discovery* 20(3): 200–216. https://doi.org/10.1038/s41573-020-00114-z. https://doi.org/10.1038/s41573-020-00114-z.

Cos, P., A. J. Vlietinck, D. V. Berghe, and L. Maes. (2006). Anti-infective potential of natural products: how to develop a stronger *in vitro* 'proof-of-concept'. *Journal of Ethnopharmacology* 106(3): 290–302. https://doi.org/10.1016/j.jep.2006.04.003.

Gertsch, J. (2009). How scientific is the science in ethnopharmacology? Historical perspectives and epistemological problems. *Journal of Ethnopharmacology* 122(2): 177–183. https://doi.org/10.1016/j.jep.2009.01.010.

Halliwell, B., and J. M. C. Gutteridge. (2015). *Free Radicals in Biology and Medicine*. New York: Oxford University Press.

World Health Organization. (2018). Noncommunicable diseases. Accessed August 1, 2021. www.who.int/news-room/fact-sheets/detail/noncommunicable-diseases.

Index

α-amylase 176
α-glucosidase 175

Antibacterial 9, 70, 71, 84, 105, 113, 119, 120, 122, 158, 160
Anti-inflammatory 73, 100, 158
Antimicrobial 12, 70, 71, 86, 119, 132, 135, 158
Antioxidant 58, 66, 67, 70, 72, 84, 114, 116, 118, 136, 145, 146, 168
Antiproliferative 74
Antiviral 71, 72

Biodiesel 2, 16
Biofuel 58

Carotenoid 104, 105, 134, 135
Cluster analysis 27, 28
Cuttings 14, 34, 41, 46, 49, 50
Cytochrome P450 25, 27, 28, 29

Diabetes mellitus 72, 134, 167, 174, 175
DNA barcoding 25, 27

Flavonoids 67, 133
Flowering 42, 44
Free radicals 70, 136, 137, 147

Genetic diversity 21–23, 25– 27, 28–33
Genomic DNA 23, 24
Genotypes 22, 27–34
Germination 14, 48, 50
Glucosinolates 3, 66, 68, 135, 160

Heavy metals 83, 93
Herbal medicines 84, 85
Hybridization 23
Hydrolysate 118, 119, 121

Isothiocyanates 66, 68, 72, 73, 86, 135

Long pods 41

Medicinal uses 9, 26, 30
Mexico 11, 42, 46, 48, 54, 57, 70, 85, 102, 103

Microsatellites 24
Molecular markers 22, 23, 26, 29, 31, 33
Moringa pygmaea 8, 10, 131
Moringa arborea 1, 4, 10, 130
Moringa concanensis 6, 10, 106, 130
Moringa drouhardii 6, 10, 12, 129
Moringa hildebrandtii 6, 10, 129
Moringa longituba 1, 7, 10, 131
Moringa ovalifolia 8, 10, 30, 129
Moringa peregrina 8, 10, 21, 29, 102, 103
Moringa rivae 8, 10, 12, 131, 147
Moringa ruspoliana 12, 131
Moringa seeds 3, 4
Moringa stenopetala 9, 11, 12, 30, 32, 129
Moringaceae 2, 3, 9, 10, 15, 26, 26, 40, 99, 113, 128

Open reading frames 25

PCR amplification 24
Peptides 113, 115, 116, 117, 118, 119
Pesticide residues 93, 94
Pests 13, 15, 50
Phenolic acids 67, 69, 72, 104, 128, 132, 133, 146, 178
Phenological 33, 40, 42
Phytochemicals 66, 99–101, 103–105, 137, 159, 160, 165, 178, 183, 195
Plant propagation 46
Pruning 40, 52, 56

Quality control 89, 93, 94, 95

Sarcorhizal trees 3, 4
Short pods 41
Slender trees 4, 128, 129, 130
Soil 13, 14, 41, 43, 45, 46, 47, 48–53
Sowing 14, 50, 46, 47, 50, 52

Tannins 54, 67, 68, 72, 128, 158, 160, 165, 176
Transplantation 46, 48, 49, 50
Tuberous shrubs 3, 4, 2, 100, 127, 129, 130

variants 41